Math for Moms and Dads

A Dictionary of Terms and Concepts . . . Just for Parents

Introduction by Suzanne Beilenson

Math Content by Catherine V. Jeremko and Colleen M. Schultz

PUBLISHING

New York

Published by Kaplan Publishing, a division of Kaplan, Inc.
1 Liberty Plaza, 24th Floor
New York, NY 10006

Printed in the United States of America

Library of Congress Cataloging-in-Publication Data has been applied for.

10 9 8 7 6 5 4 3 2 1

ISBN-13: 978-1-4277-9819-0

Contents

About the Authors

After graduating from Harvard College, **Suzanne Beilenson** taught mathematics at two New York private schools—Rye Country Day School and The Spence School. She has worked with students at all levels of math ability, in classes ranging from pre-algebra to calculus. Since then, her math background has become as helpful at home as it had been in the classroom. She is the mother of two and is on call for math-related homework questions every day.

Catherine V. Jeremko is a certified secondary mathematics teacher in New York State. She is the author of mathematics test prep and review materials. She currently teaches seventh-grade mathematics at Vestal Middle School in Vestal, New York, and also serves as a regional math coach. Ms. Jeremko is a teacher trainer for both mathematical pedagogy and the use of technology in the classroom. She resides in Apalachin, New York, with her three daughters.

Colleen M. Schultz is certified in both secondary mathematics and elementary education in New York State. She is the author of mathematics test prep and review materials. She currently teaches eighth-grade mathematics at Vestal Middle School in Vestal, New York, where she has also served as a teacher mentor for the Vestal School District. Mrs. Schultz is a teacher trainer for both mathematical pedagogy and the use of technology in the classroom, as well as a regional math coach. She resides in Binghamton, New York, with her husband and three children.

Introduction

Thirty years ago or so, you sat in math class, staring at the blackboard as your teacher derived the Pythagorean theorem. If you break out into a cold sweat at the memory, you are not alone. Those classroom experiences left many adults math-a-phobic, avoiding all math-related endeavors in the ensuing years. Therefore, it is one of life's little ironies that for all you may have since accomplished—college, the job market, parenthood—you now find yourself back in math class. Only this time, you aren't worrying about your grade . . . but rather your child's success.

As nerve-racking as it may be to watch your child struggle, the real anxiety stems from not knowing how to help. That's where *Math for Moms and Dads* can make a big difference. As a parent and math teacher, I have sat on both sides of the homework assignment. I know the stress of watching my own kids grapple with math concepts. Yet having worked with students of all abilities, I also know that any child can learn, improve, and earn higher grades in math. *Math for Moms and Dads* walks you through the techniques and strategies that you, as a parent, can employ to help your child succeed.

Yes, you really can help. However, before tackling irrational numbers or exponents, the road to success begins with the right attitude. Make math a positive experience for your child. For too many students, math class is fraught with confusion, stress, and frustration—feelings that can overwhelm the learning process. By alleviating the negative feelings that your child associates with math, you will create room for change to happen. There are several ways parents can help lessen the anxiety.

Too often, stepping into math class feels like landing in a foreign country. When language is unfamiliar, reading the signs or even having a conversation becomes arduous. I have had students panic when they see the word *integer* or *identity* or *inverse*. Why? Because it is virtually impossible to answer a question correctly if you're not sure what's being asked. Unlike Spanish or French class, where students must memorize vocabulary, math teachers rarely drill students on the meaning of math terms.

Therefore, the first step toward making your child more comfortable and confident in math class is to help improve his or her math fluency. Use the vocabulary section of *Math for Moms and Dads* to review and clarify all of the terminology you need to know to speak "math" with your child. The more often you use these math terms with your child, the more easily fluency will come for both of you.

A second approach for making math a positive experience is to model the behavior you want to see in your child. Too often, I hear parents saying "I can't do math," or "I hate math." Try not to make your problem your child's problem. If you, as your child's role model, "can't do math," then the first time your child encounters difficulties in math class, the response will be "I can't do math, either." The primary purpose of *Math for Moms and Dads* is to help build your confidence, because a parent's own math phobia single-handedly creates the biggest obstacle to kids' success.

Third, cheer your child on—especially when the going gets rough. There's a big difference between finding math difficult and not being able to do it. I have seen the greatest confidence boosts in students who do not have a natural facility with numbers but who have had to work really hard in math class. Every time students learn that they have the capacity to take on a challenge and win, they learn a giant life lesson and confidence soars. That is the great value in learning how to do something hard. The next time a challenge—academic or otherwise—presents itself, your child will be more likely to say, "I can do this. If I learned how to use the quadratic formula, I can do this, too." Keep your child motivated during the hard moments by always applauding efforts. It helps as much in math class as it does on the soccer field!

Finally, help your child see the connections between math and everyday life. The less abstract math is, the more manageable it becomes. As a parent, I always try to illustrate for my kids how math surrounds us everywhere in our daily lives. In my house, baking cupcakes is proof of the commutative law of multiplication: three rows of four cupcakes equal four rows of three cupcakes. Going to Starbucks is a trip on the coordinate plane: our house is the center of the universe at (0, 0), and coffee is three blocks up and two blocks left at (3, –2). Putting money in a savings account is a geometric progression that really pays off! Soup cans are cylinders, cereal boxes are prisms, and basketballs are spheres.

The next time you go out for dinner, let your child check the bill for errors or figure out a 15 percent tip. Ask him to double the recipe quantities for Aunt Martha's famous chocolate cake. Moreover, if your child ever asks "How much longer?" on a road trip, simply state the time you left, the distance of your destination, and your average speed.

The more you make math an evident part of your child's life, the less math class will seem like a foreign country. Help your child have a great experience and feel comfortable by learning the language, being a positive role model, and making math a tangible part of life. You will both reap the rewards.

As important as it is for math to be a positive experience, establishing good habits will take your child further down the road to success. These good habits include excellent computation, showing all your work every time, and a disciplined homework routine.

Good habits matter in math. While mathematics is theoretically based, school-level math is rather concrete. It is a contained world. The problems always work out. They always rely on information in the chapter. There is always a right answer. It is generally far more mechanical than theoretical. Students can earn good grades in math by mastering procedures instead of being intellectually creative. That mastery requires good habits.

In fact, I have seen kids with a tenuous grasp on a particular concept still ace a test because they have become master technicians. As a parent, you can help your child do the same.

The Good Habit of Excellent Computation

You would never expect a middle school student to read Aristotle or Tolstoy, but you prepare them for the task by giving them a foundation in grammar, vocabulary, and literary concepts. The same holds true for math. Higher math—from algebra to calculus—requires a strong foundation, and students have to acquire certain skills before meeting the challenge. The building blocks are not conceptual. They are computational. They are arithmetic, fractions, decimals, exponents, and square roots.

I have been shocked at how many students label themselves "bad at math" when actually they just have weak computational skills. Even though a student may understand the math concept in question, errors in computation lead to wrong answers. Enough wrong answers, and you have a frustrated and floundering math student.

The good news is that practice does indeed make perfect or, at least, substantially better. The bad news is that a calculator is not the answer. Do not misunderstand. I love a good calculator, and it is a formidable asset when used appropriately. However, if I am going to bet, I always put my money on the kid who knows the times table cold over the kid with the graphing calculator for some simple reasons:

- The calculator is only as good as the person using it. Make a mistake entering a number, and you will get the wrong answer.
- When it comes to basic computation, the brain is faster than a calculator.
- Using a calculator for computation interrupts your train of thought.
- Most importantly, building facility with numbers builds pattern recognition.

Therefore, drill your kids. Drill them in the car, at the breakfast table, or on their cell phones. Keep at your child until he can answer without *any* hesitation that 7 times 8 equals 56 or 12 times 6 equals 72. They will thank you when they have to factor the quadratic expression $x^2 + 15x + 56$. Because when you know that the factors of 56 are 7 and 8, the answer comes quickly and easily: $(x + 7)(x + 8)$.

If you find that your child is weak when it comes to fractions, exponents, or division, ask the teacher for a review sheet or download one online. Buy flashcards. The better your child's computation, the easier the math will be—and the higher your child's grade.

The Good Habit of Showing All Your Work

Over and over again, I have witnessed Aesop's fable of the tortoise and the hare come to life in the math classroom. Whether it's out of dislike or impatience, some kids approach math as if it were a sprint—the faster, the better. Rather than systematically writing out every line of their work, these "hares" only document what they feel is necessary. Unfortunately, it is far too common to veer off course when one is moving at breakneck speed. Invariably, notation is lost, terms are dropped, and the resulting answer is incorrect—particularly when the stress of test taking enters the mix.

I had a very bright student who simply refused to write out every line of her work. Every time she was solving a quadratic equation, she would drop the '= 0' at the end of each line because, in her opinion, it was unnecessary. The '= 0' wasn't changing, so why repeat it? Sure enough, she almost always stopped working after factoring the expression. She forgot to set each factor equal to zero and actually solve for x.

In the end, students make fewer mistakes when showing all of their work, and they finish the assignment or the test faster because they do not have to retrace their steps. The moral of the story: the tortoise wins the race in the end.

Still, there's a second, even better reason for showing all your work every time. Documenting the work means that your child can easily check it for mistakes—and so can the teacher! From a confidence-building perspective, showing all your work is crucial for kids. When students can review and find their errors, making mistakes stops being scary, and they become better independent learners. An incorrect answer is no longer a dead end but a starting point.

At the test-taking level, showing all the work allows math teachers to give partial credit. If your child makes a computational error along the way but the rest of the math is correct, she will earn more points. Math is not only about getting the right answer. It is also about the process.

The next time your child complains that showing all the work is a waste of time, remind him that showing *all* the work *every* time increases precision and attention to detail, and building these skills will pay off in fewer mistakes and higher grades.

The Good Habit of a Disciplined Homework Routine

Homework may be the bane of your child's (and your) existence, but a disciplined homework routine yields significant improvements in comprehension and test scores. Do not let your child waste this daily golden opportunity.

To begin, make math homework the priority. It comes first before all other assignments. After a full day of school followed by extracurricular activities, kids are tired, but they are only going to get more fatigued as the day wears on. Take advantage of their energy levels—both mental and physical—earlier in the afternoon or evening and get the math homework out of the way first.

Next, eliminate distractions. Unplug the computer, the television, the music, and the cell phone. Math homework gets done faster without the diversions, and equally importantly, the homework setting should mimic the testing setting. I highly doubt your child has ever taken a test with music playing in the background.

Third, find the right room for doing homework. The kitchen table may be too busy. The family room usually has the television. The bedroom can be a text-messaging center. If the usual haunts are not working, try the dining room if you have one. There's plenty of room to spread out at the table, and there are limited distractions. Plus, you can keep a watchful eye on any covert cell phone use!

Finally, whatever else you do, do not hover during homework. Be on call if your child has a question, but sitting next to your child during homework only increases her reliance on you—both emotionally and academically. You won't be there during the test, so your child has to develop confidence in her own abilities.

Yet there is a more vital reason for your child to do homework without your help. Homework is a gold mine of information. There among eraser shavings and creased pages, all is revealed. My most effective method for helping students improve in math has been to collect homework—*at the start of class*. I do not want to see homework that has been corrected, because uncorrected homework contains all the information about why a student is struggling. Misconceptions, weak computational skills, and even a poor study environment are all exposed in homework. As a parent, let your children do their homework to the greatest extent possible on their own. Only then review the work if need be.

"But my child will make mistakes!" is a parental refrain I have heard over the years. That's okay. Everyone learns by making mistakes, and homework needs to be a "safe place" for making errors. When you review homework, you will certainly find mistakes. That's great! It really is. When you find mistakes, you find the root cause of the problem. Then use that knowledge to target where your child needs help.

Your child is going to spend many hours doing math homework. Make that investment pay off. Schedule math homework first and ensure your child does it in an environment truly conducive to learning. Only get involved as you need to be and use mistakes as a map for getting back on track.

A disciplined homework routine is one habit you want your child to develop. Combined with the good habits of strong computational skills and showing all the work—every time—your child will see gains in grades as well as in confidence.

Finally, if helping your child with math feels like a daunting proposition, remember you are not alone. You have a highly educated resource at your disposal: your child's teacher. Because establishing a positive relationship with the teacher is hugely beneficial, chapter 8 of *Math for Moms and Dads* is all about communicating effectively with teachers.

Now brace yourself here. Some parents—more than you think—do not have great relationships with their children's teachers. In trying to elicit a teacher's help, the approaches parents take can often backfire. Being a parent myself, I am loathe to criticize moms and dads, particularly those who care enough to worry about their child's math abilities! Just like you, I have lain awake at night worrying about my children, so when I need to talk to a teacher, my emotions are usually running high and I am sleep deprived to boot. In that parent-teacher meeting, I am discussing *the most important kid in the world*. Understandably, I may not necessarily be at my coolest and most collected.

Yet as much as I empathize with parents, I have been on the receiving end of those parent-teacher interactions, too. It is not always pleasant. Worse, it is not always productive. Luckily, it does not have to be that way. Keep in mind the following three points, and you will set the stage for a satisfying, productive relationship with your child's teacher:

1. Treat teachers with the respect they deserve. They have a tough job with a huge hidden workload, and compared to other professions, they are underpaid for their level of education.
2. No one enjoys being criticized. As much as you may think a teacher is at fault . . . because the test was too hard, the homework was too long, or the instruction was insufficient . . . there is a high likelihood that some or more responsibility rests on your child's shoulders. Before you judge, ask yourself how you would respond if the teacher criticized your parenting.

3. Listen–at least as much as you talk. Too often, we want to "tell" teachers what the problem is. You will get more out of any conversation when you ask the teacher for his opinions and recommendations. Sometimes, we don't like what we hear, but it's always worth considering. The teacher is a far more objective observer of our children than we are.

It's a good rule of thumb to model your relationship with a teacher after the one you have with your pediatrician. We don't show up unannounced at a doctor's office. We call first and make an appointment. We don't tell the doctor how to do her job. We listen to and trust the doctor's advice.

We also don't wait until our child has a 105° fever to call the pediatrician. We contact the doctor at the first signs of illness. Use the same standard with your child's teacher. If you think there's a problem, schedule an appointment sooner rather than later. Stay on top of your child's progress. Monitor homework, and if necessary, ask the teacher to inform you of test dates and what grades your child received. Waiting for a report card to find out if there's a problem is not effective. I had a student whose father came in at the end of the year and told me that his son deserved a higher grade than the B- on his report card. With every quiz, test, and exam behind us, I could do nothing at that point. The subsequent conversation would have been very different if the father had come to me earlier in the term.

You and your child's teacher have the same goal. You both want your child to succeed. If you respect and listen to your child's teacher, he will become a valuable member of the team. Your teacher is the one writing and grading the tests. That alone should make it worth the effort to establish a positive relationship!

As parents, we are our children's first teachers. The bigger our kids get, the bigger the challenges they face, and if math class proves challenging,

Math for Moms and Dads will help you support your child. You do not need to be a math teacher to ease your child's way. *Math for Moms and Dads* walks you through the basic rules of math, direction decoding, appropriate uses for a calculator, and problem-solving strategies. So sharpen your pencils, and by the time you finish reading, you may just surprise yourself. Math might seem more intelligible and accessible than the last time you laid eyes on the Pythagorean theorem—and hopefully, math class will be more fun the second time around.

Chapter 1

How to Use This Book and When to Use Your Calculator

The best part about this guide is that its information applies to the student learning arithmetic, the student tackling algebra for the first time, the student beginning high school geometry, and beyond. Staring math in the face and breaking it down into easily digestible parts is a skill you, as a parent, can help your child with at any level. Any sections of the book that talk about specific vocabulary or concepts that are not relevant now will definitely come up a few years down the line as your child advances through the grades.

Where Do I Go?

Each chapter of *Math for Moms and Dads* focuses on a different aspect of math. To get the most out of this text, read it straight through for a general overview and then refer to it throughout your child's scholastic math career as situations arise.

Chapter 2: Math Vocabulary contains a list of vocabulary terms that often make math more intimidating than it has to be. Refer to this chapter when you come across a term you or your child does not recognize—most likely, it can be explained in easy-to-understand language. Translating is often half the battle.

You should refer to Chapter 3: The Rules You Should Know for an overview of the basic math rules needed to tackle even the easiest math problems. These rules apply to everything from basic arithmetic to the most complex algebra. Your child should know the concepts in this chapter and eventually master them. If you're familiar with them, too, you can both approach any problem with a solid base of knowledge.

Chapters 4: Direction Decoding and 5: Little Pieces Lead to Big Problems begin teaching you how to help your child approach specific problems in homework and tests. Chapter 4 helps you answer the most important question: "What *exactly* is this question asking here?" Chapter 5 explains how even difficult problems can be broken down bit by bit into simple, easy pieces. This chapter mirrors the overall theme of this book—it's always possible to take something big and intimidating and turn it into something easy and manageable!

Chapters 6: Study Strategies and 7: "When Will I Use This, Anyway?" offer advice on how you can help make math more digestible for your child in general. The former chapter offers strategies for creating ideal study, homework, and test-taking habits. The latter chapter gives you concrete things to say when your frustrated child inevitably wonders, "Why do I need this stuff, anyway?"

Finally, Chapter 8: Parent-Teacher Communication is your resource on how best to communicate with your child's teacher to create a supportive learning system that will lead to math success.

Is My Child's Learning on Track?

The National Council of Teachers of Mathematics (NCTM) has a list of set topics that students should master for each grade to meet state standards. Refer to this book's appendix for guidance on the levels your child should accomplish at each grade. Keep in mind that this listing is only a general guide summarized from national standards, not a curriculum. For more detailed requirements and specific state standards, always ask your child's teacher or guidance counselor, who should be able to tell you your state's board of education standards or where to find them.

The appendix also features a "What Kind of Learner Is Your Child?" quiz you can take with your child to come up with effective study strategies, as well as a handy conversion table of measurement and other values that come up again and again—not only in math problems but also in real life.

The Calculator: How Important Is It?

As you learned in the introduction, calculators are never the answer to understanding and acing math. The bottom line is that your brain is the most important tool in solving math problems. Always remind your child of this. The calculator is only as effective as the person pushing the buttons.

That said, when used in the correct context, the calculator *can* be your friend. In math classrooms of the past, use of a calculator was frowned upon. In today's math studies, there are problems that require the use of this tool. It is important that your child understand how to use a calculator effectively.

Many current calculators on the market have a two-line display. This is very helpful, because it shows the keystrokes typed in. If your child's calculator has this feature, he can look at the display to double-check what was keyed in. If the calculator does not display the keystrokes, then it is necessary to retype any sequence to be sure all of the numbers and operators were entered correctly.

Calculator Features

In addition to the basic buttons for addition, subtraction, multiplication, and division, many current calculators have additional features.

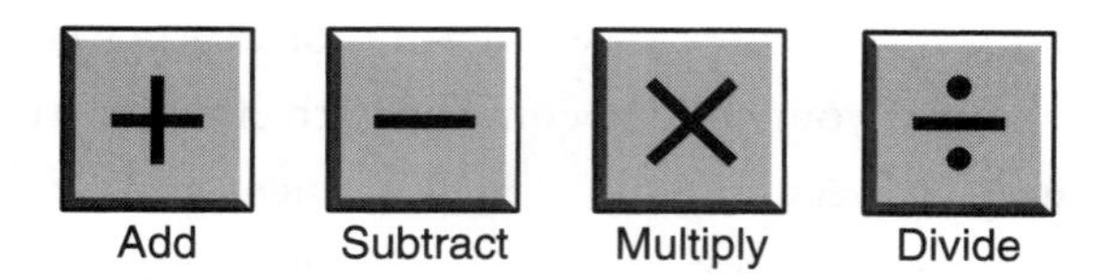

Order of Operations

Some calculators do not abide by the correct order of operations (this term is explained further in chapter 2). You can test your calculator with a simple expression, such as $4 + 10 \div 2$. If your calculator shows the answer as 9, it does follow order of operations. If it displays a 7, it does not.

If your calculator has parentheses keys, then it most likely follows the correct order of operations.

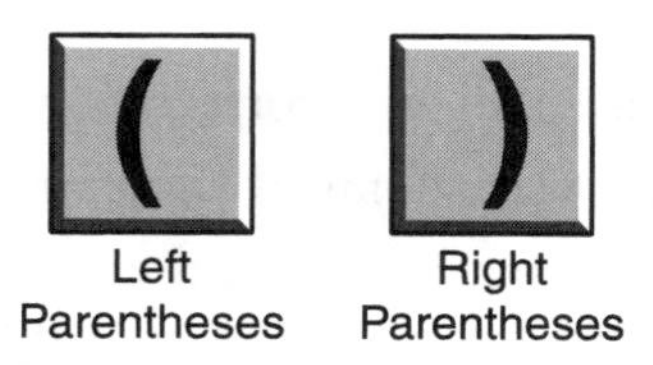

Negative Numbers

Many calculators have a separate key to enter negative numbers. It is not the subtraction key. Look over your calculator to see if this key is present. If you have this function, practice with it. Some calculators have you enter the negative sign first (the most natural way), while others require you to enter the number first and then the sign key.

Powers and Roots

Most calculators have a special key used to square a number, and it generally looks like the key on the left in the figure below. This key is present because squaring is such a commonly required operation throughout high school mathematics. The opposite of the squaring operation is the square root operation. Most calculators have a key to do the square root function. This key is also shown below.

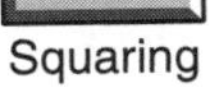

Square Root

Locate these keys, the square and the square root, on your calculator. Test your calculator to determine the correct order to use when squaring and taking a square root. Almost always, to square a number, you enter the number first and then press the square key. For square roots, it is more calculator brand-specific. On some calculators, you enter the number followed by the square root key; on others, you press the square root key first and then enter the number.

Exponents

There is also an exponent key on most calculators. This is used for exponents larger than two (squaring). The exponent key typically looks like one of the two keys shown here.

Two examples of types of Exponent Keys

With either key, you enter the base number, then one of the keys shown above, and finally the exponent value.

The Π Constant Key

Usually, calculators come equipped with a key to give the constant value of pi (π). This constant is used when measuring the circumference or area of circles. Sometimes, your child's teacher may ask students to use the approximation of 3.14 for the value of π. In this case, you would not use the special key. The π key gives a more accurate approximation, such as 3.141592654. Pay attention to what the teacher directs the students to use; sometimes your child will be directed to use the special key. Because of rounding errors, answers can be different depending on which value of π is used. Other times, the teacher may direct the students to leave an answer in terms of π. In these cases, do not use the calculator to approximate π and do not use 3.14 for the value of π. Just leave π in your answer.

Fraction Arithmetic

Many calculators have fraction features. Your child's teacher may not allow the calculator when learning fractional computation, but there may be times when a calculator can be used for these operations. Even if your child is learning fraction arithmetic, he can still check answers with the calculator.

The fraction keys generally look like one of the keys shown below. You enter the numerator, then press the fraction key, and then enter the denominator.

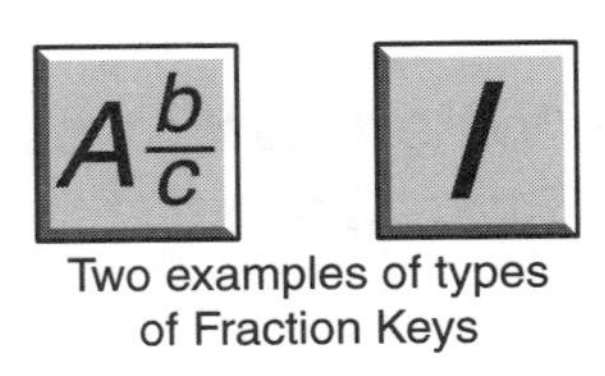
Two examples of types of Fraction Keys

If one of these keys is present on your calculator, you can add, subtract, multiply, or divide fractions.

The Graphing Calculator

A more elaborate and expensive calculator is the graphing calculator. Many high schools are now requiring their students to purchase this type of calculator. In addition to having all of the features already described, these calculators will also graph functions on a coordinate grid. The graphing calculators have a much larger screen for viewing graphs and lists of numbers. They also have many more functions and built-in formulas that your child will learn and need in high school.

Generally, the graphing feature is fairly straightforward. There is a key used to enter in a linear function, which may look like the key displayed here. In addition, there is a variable key to enter the variable *x* and a graph key to display the graph.

Key to set up the equation to graph
Variable Key
Graph Key

To graph a line, for example, first press the Y= key, then enter in a linear function, such as $3x + 5$. Use the variable key to type the *x* in the expression $3x + 5$. Then, press the graph key and the graph of the function appears on the screen.

Determine When It Is Appropriate to Use a Calculator

There are situations when it is definitely advantageous to use a calculator. Sometimes, the teacher will allow the use of a calculator when the goal of the learning is a math concept that does not exclusively center on computation, such as when solving difficult equations or learning probability concepts.

Some examples when a calculator is most useful are as follows:

- Problems calling for the use of the π key when computing circumference or area of a circle, to get accurate results
- Problems involving finding square roots, such as those that use the Pythagorean theorem
- Problems involving a lot of division in which learning the long division method is not the main objective.

There are also situations when it is more efficient to use your brain instead of a calculator. Following are some example:

- When multiplying by powers of 10, such as 10, 100, or 1,000, it is quicker just to move the decimal point the appropriate number of places to the right:
    - To multiply by 10, move the decimal point one place to the right.
    - To multiply by 100, move the decimal point two places to the right.
    - To multiply by 1,000, move the decimal point three places to the right.

- When dividing by powers of 10, such as 10, 100, or 1,000, it is quicker just to move the decimal point the appropriate number of places to the left:
    - To divide by 10, move the decimal point one place to the left.
    - To divide by 100, move the decimal point two places to the left.
    - To divide by 1,000, move the decimal point three places to the left.
- When you can estimate an answer, especially in a multiple-choice question, using one's head instead of the calculator is often faster.
- Many times, when you need to order numbers, such as from least to greatest, you can use estimation and comparisons.

What Have We Learned About Calculators?

- Your brain is the most important tool when solving math problems.
- Check to see if your calculator follows the order of operations.
- Check for the negative key, which is different from the subtraction key.
- Locate the squaring, square root, and exponent keys and learn how they work.
- Locate the π key on your calculator.
- Determine whether your calculator does fractional arithmetic and learn how to use this feature.

- Graphing calculators can graph functions.
- In certain situations, using a calculator is advantageous; in others, it is more efficient to use your brain. Learn these distinctions.

Now that you know your way around this guide and have the basic calculator rules under your belt, you're ready to dive into the world of *Math for Moms and Dads*. Good luck, and have fun!

Chapter 2

Mathematics Vocabulary

Mathematics has a vocabulary all its own, and often students are confused or intimidated by complicated sounding terms. Very often, the language of math is used to ask something that appears difficult but is pretty simple. For example, your child may encounter a problem that asks for "the product of six and ten" when all she is really being asked is to multiply 6 times 10.

Help ease your child's mind by reminding him that often mathematical terms are not as complicated as they seem and many can be broken down into pieces that are easier to understand. When appropriate, it is helpful to relate them to the everyday meanings of the terms so that your child can make a connection to his own life.

This chapter of the book presents a vocabulary list of mathematical building blocks. Use this list as a reference when your child encounters an unknown or unclear word on a project or homework. Also, a section of math synonyms clarifies the differences between commonly misunderstood terms. A comparison of these terms reveals that while they may be close in meaning, they have critical differences. Finally, the last section in this chapter presents you, the parent, with tips and strategies to use when helping your child attain a solid math vocabulary.

Vocabulary: The Basic Building Blocks

Absolute Value: The positive value of a number or expression. The symbol for absolute value is $|x|$, which reads "the absolute value of x." Think of absolute value as the distance that a number is from zero on a number line; distance is always a positive value.

Acute Angle: An angle whose measure is greater than 0 degrees and less than 90 degrees.

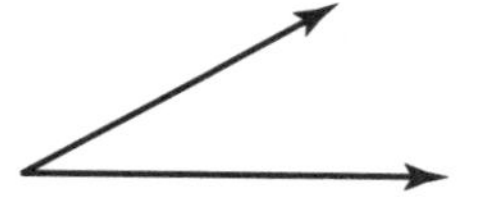

Angle: A geometric figure consisting of two rays that share a common endpoint. The common endpoint is called the vertex. The vertex of $\angle ABC$ is point B. Angles are measured with a protractor, and the units are expressed in degrees.

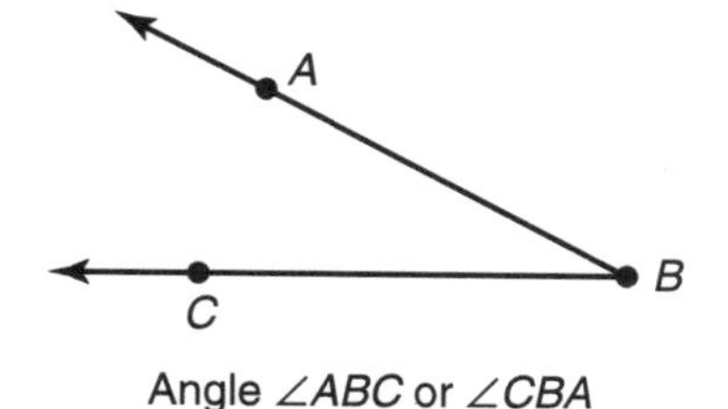

Angle $\angle ABC$ or $\angle CBA$

Area: The amount of square units it takes to cover a geometric figure. Area is the inside of a two-dimensional object, and the units are expressed in square units.

Circle: A two-dimensional geometric figure consisting of all of the points that are the same distance from a fixed point, called the *center*. The center is not a point of the circle itself; it just defines the circle.

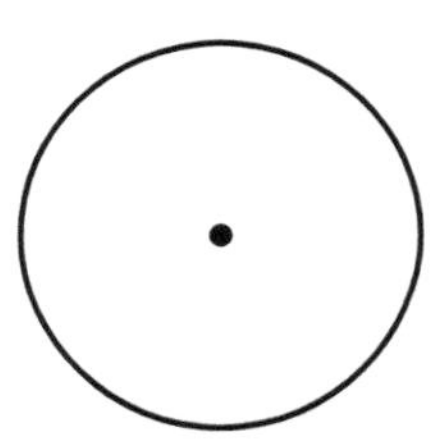

Circumference: The distance around a circle. The formula to find the circumference is the product of the constant term π and the diameter: $C = \pi d$.

Combinations: The number of different groupings of a set of objects, without regard to the order of the objects in the group. The formula for n different objects, taken r at a time, is ${}_nC_r = \frac{n!}{r!(n-r)!}$.

Complementary Angles: A pair of angles whose sum measures 90°. In the following figure, angle 1 and angle 2 are complementary.

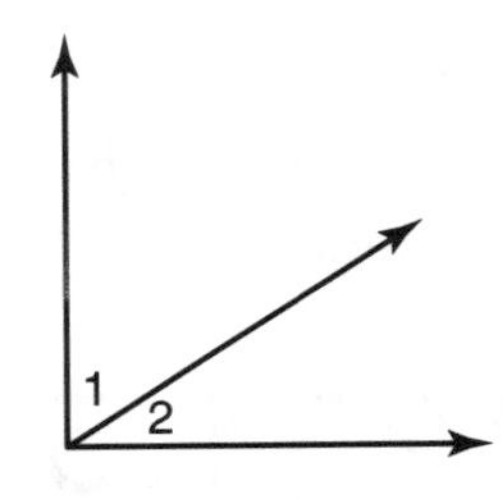

Composite Numbers: Counting numbers {1, 2, 3, 4 . . .} that have factors other than 1 and the number itself.

Congruent: Geometric figures that have the same size and shape. All corresponding angles and sides have the same measure.

Constant: A number, sometimes represented as a symbol, that does not change value.

Coordinate Plane: A two-dimensional flat grid, defined by the intersection of a horizontal line (the *x*-axis) and a vertical line (the *y*-axis).

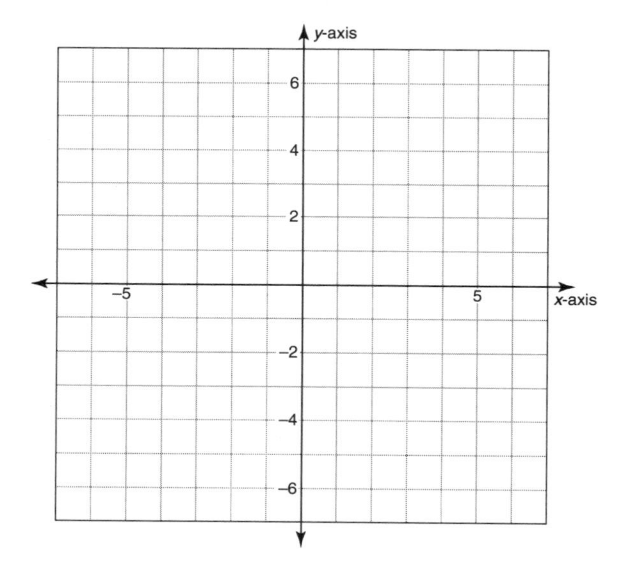

Decimal: A number expressed in the base 10 number system, using place value. Decimal numbers have a decimal point. The digits to the left of the point are in the ones, tens, hundreds, thousands (and so on infinitively) place. Digits to the right of the decimal point are in tenths, hundredths, thousandths (and so on) place.

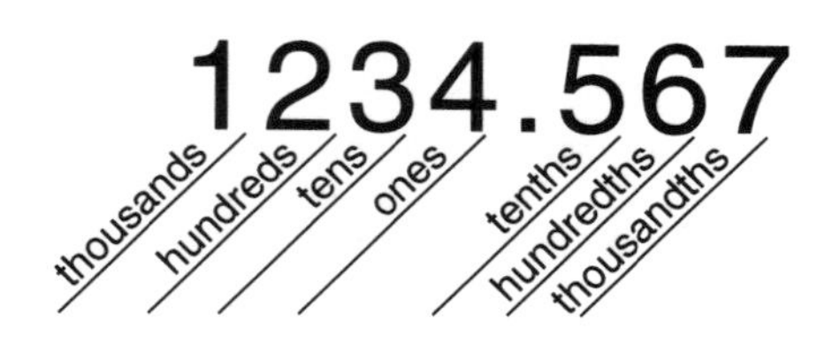

Denominator: The bottom number or expression of a fraction that often represents the whole portion, when comparing part of a whole. For example, in the fraction $\frac{2}{3}$, 3 is the denominator.

Diagonal: A line segment whose endpoints are any nonadjacent (not next to each other) vertices. In the following figure, line segment *AC* is a diagonal of rectangle *ABCD*.

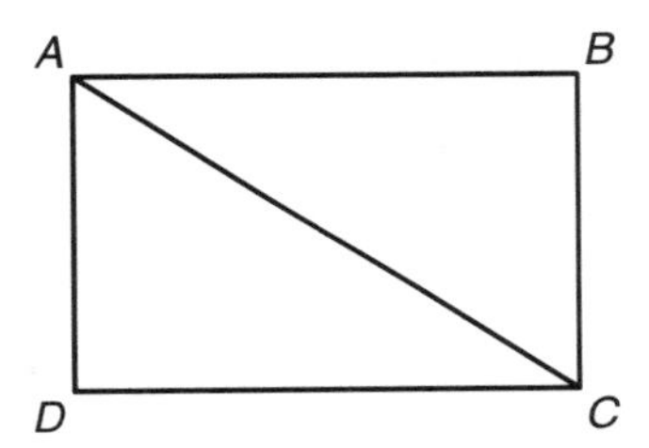

Diameter: A line segment with endpoints on the circle that passes through the center of the circle. In the following figure, line segment AB is a diameter of the circle with center C.

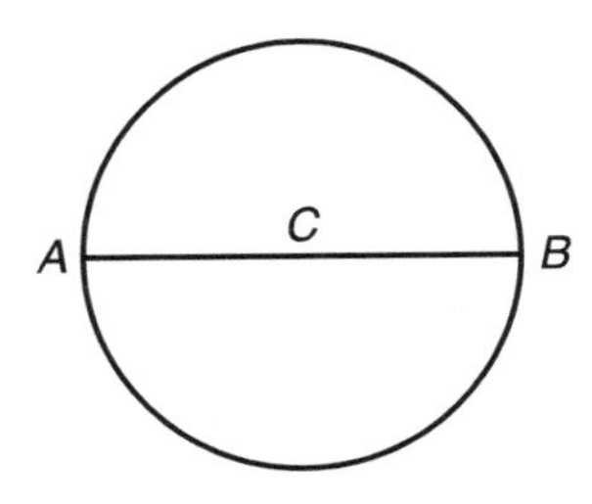

Equation: A mathematical sentence that states that two expressions are equal. Usually, equations contain a variable.

Exponent: A number, shown as a superscript (above and to the right of a number or variable), that tells how many times a base value is multiplied by itself. For example, in the statement $5^3 = 5 \times 5 \times 5$, the 3 is the exponent.

Expression: A mathematical phrase that contains terms, which are numbers or variables, and operations, such as addition or subtraction.

Factor: A positive integer that divides into a number with no remainder. The number 1, and the factor itself are considered to be factors of a number.

Fraction: A number that is used to express a ratio, often as part of a whole. A fraction is made up of a numerator and a denominator. For example, in the fraction $\frac{5}{16}$, 5 is the numerator and 16 is the denominator. Every rational number can be expressed as a fraction.

Improper Fraction: A fraction whose numerator is larger than its denominator.

Integer: The whole numbers {0, 1, 2, 3 . . .} and their opposites {. . . −3, −2, −1, 0, 1, 2, 3 . . .}.

Mean: The average of a set of numerical data. The sum of the set of data, divided by the number of data elements. The mean is referred to as one of the *measures of central tendency* used to describe a set of data.

Median: The middle value when a set of numbers is ordered from least to greatest. If there are two numbers in the middle, then it is the mean of the two middle values. The median is referred to as one of the *measures of central tendency* used to describe a set of data.

Mode: The value that appears most often in a data set. There can be no mode, one mode, or two modes (a bimodal set).

Multiple: A whole number that another whole number divides into without remainder. For example, some multiples of 6 are 6, 12, 18, and 24.

Numerator: The top number or expression in a fraction. Often, it represents the part when comparing a part to a whole. For example, in the fraction $\frac{5}{7}$, 5 is the numerator.

Obtuse Angle: An angle whose measure is greater than 90 degrees and less than 180 degrees.

Ordered Pair: Two numbers, written as (x, y), that indicate a location on a coordinate plane. The first number is the distance to the left or right of the y-axis, or the origin, and the second number is the distance above or below the x-axis, or the origin.

Origin: The intersection of the x-axis and the y-axis on a coordinate plane.

Percent: A ratio used to compare a number to 100. Percents can be written with the percent symbol (56%), as a decimal (0.56), or as a fraction $\left(\frac{56}{100}\right)$.

Perimeter: The distance around a geometric shape. Perimeter is found by adding the lengths of all of the sides of a geometric polygon.

Permutations: The number of possible arrangements of a set of objects in which the order of the objects matters. The number of permutations of n objects taken r at a time is ${}_nP_r = \frac{n!}{(n-r)!}$.

Pi (π): The constant ratio of the circumference to the diameter of a circle. The value of π never changes, and it is an irrational number. The approximate value is 3.14156 . . . Often in math classes, students are asked to use 3.14 for the value of π, or they are asked to use the π key on a calculator, which carries the value out at least five decimal places. Answers to problems involving π will vary depending on which approximation is used.

Prime: A positive integer that has no factors other than 1 and itself. Prime is the opposite of composite. Zero and 1 are neither prime nor composite. There is no pattern to the prime numbers, nor is there a formula to generate all of them. Students must either memorize or test to see if a number is prime. The only even number that is prime is 2. The first seven prime numbers are 2, 3, 5, 7, 11, 13, and 17. There is an infinite number of prime numbers.

Probability: A ratio that indicates the number of ways an event can occur compared to the total number of possible outcomes. Probability can be expressed as a fraction, decimal, or percent. The notation for the probability of event E is $P(E)$.

Proper Fraction: A fraction whose numerator is less than its denominator.

Quadrilateral: A four-sided closed geometric figure.

Radius: A line segment whose endpoints are the center of the circle and any point on the circle. The radius length is one half the length of the diameter. In the following figure, line segment AC is a radius of circle with center C.

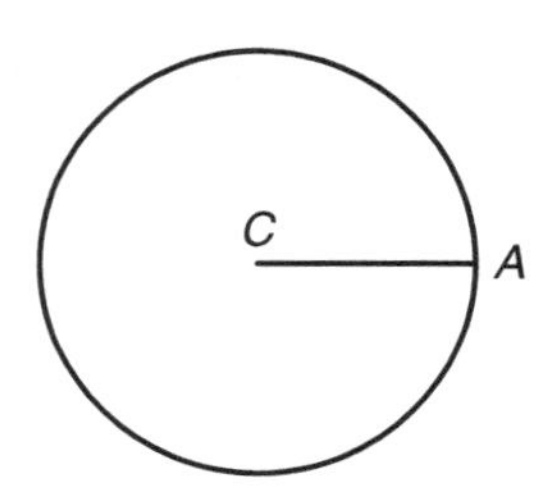

Range: The difference between the highest and lowest values in a data set. This is called one of the *measures of central tendency* and indicates how closely spaced the data is in a set.

Rate: A ratio that compares two unlike items, such as miles and hours or dollars and pounds. A unit rate is a rate in which the denominator is 1.

Ratio: A comparison of two numbers or expressions. A ratio can be written as $a{:}b$, a to b, or $\frac{a}{b}$.

Right Angle: An angle whose degree measure is 90 degrees.

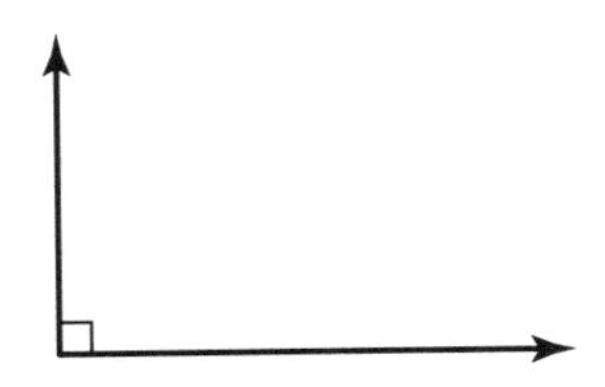

Similar: Geometric figures with the same shape but not necessarily the same size. All angles are congruent, and the corresponding sides are in proportion.

Simplify: To convert a number or expression into its simplest form. To simplify a fraction means to convert it to its lowest terms. To simplify an expression means to perform all orders of operations and combine any like terms so that it is shown in its simplest form.

Solve: To perform simplification and opposite operations on an equation to isolate the variable on one side of an equation.

Square: The operation of multiplying a number with itself. It is also referred to as taking a number to the *second power*.

Square Root: This is the opposite operation to squaring. The square root of a number is the value that, when multiplied by itself, equals the number. An algebraic description: When $a^2 = b$, a is called the *square root of b.* The symbol for square root is $\sqrt{}$. That is, when $a^2 = b$, $a = \sqrt{b}$.

Supplementary Angles: A pair of angles whose sum equals 180 degrees. When these angles are adjacent (share a side) they form a line and are also called a *linear pair*. In the following figure, angle 1 and angle 2 are supplementary and form a linear pair.

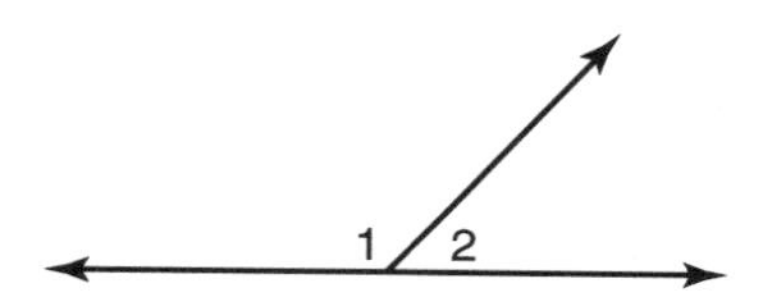

Surface Area: The sum of the areas of all of the faces of a three-dimensional figure.

Triangle: A geometric figure with three sides; this is one of the most important of the geometric figures. The right triangle, which contains one right angle, is used in both measurements and in trigonometry.

Variable: An unknown value, usually represented by a letter of the alphabet.

Volume: The number of cubic units it takes to fill a three-dimensional solid.

***x*-axis:** The horizontal number line that, with the *y*-axis, forms a coordinate plane.

***y*-axis:** The vertical number line that, with the *x*-axis, forms a coordinate plane.

Commonly Confused Terms

An area that causes a lot of confusion in math, is commonly confused terms. These are words whose meanings in mathematics are similar, but each of the words has its own characteristics that makes different. This section lists common pairs, and sometimes triplets, that your child will encounter during math studies. It also compares and contrasts them to highlight the subtleties. To make them easier to locate, the pairs are grouped into the categories Number and Operations, Algebra, Geometry and Measurement, and Probability and Statistics.

Numbers and Operations

Sum and Product

Sum is a key word in math for addition or plus. In a word problem, the student may be asked to find the "sum of five and four." In this case, you would add $5 + 4 = 9$ to find the answer. The sum is the result, which is 9.

The key word *product* is associated with multiplication. Specifically, the product is the result of multiplying values together. For example, the product of six and three is $6 \times 3 = 18$. The actual product is 18.

Students often confuse these two terms, multiplying when they see the word *sum* and adding when they see the word *product*. One way to help keep them straight is to remember that the word *add* has three letters, as does the word *sum*. The word *product* is longer, as is the term *multiply*.

+ = add = sum
× = product = multiply

Commutative Property and Associative Property

Two of the most frequently confused properties studied in math, each shows that you can take an expression and change it a certain way but the result remains the same. However, each property changes the expression in different ways.

The *commutative property* states that the order of the numbers being added or multiplied can be switched but the sum or product stays the same. For example, $3 + 4 = 7$ and $4 + 3 = 7$. In multiplication, the value of $5 \times 8 = 40$ and $8 \times 5 = 40$. The order can be changed, but the result is the same.

Because of the *associative property*, the grouping of the numbers being added or multiplied changes, but again, the result is the same. When adding $3 + (4 + 5)$ using the correct order of operations, what is in the parentheses $(4 + 5)$ is added first, and then the sum (9) is added to the number outside of the parentheses (3) to get a final result of 12. Without changing the order but changing the grouping, the problem can also be written as $(3 + 4) + 5$. Using correct order of operations, adding inside the parentheses first makes the expression $7 + 5$, and then that is added, yielding the same final result of 12. Changing the grouping did not change the result.

To remember the difference between the two properties, use the first few letters in each term. For the commutative property, take the first two letters *co* to stand for "**c**hanges **o**rder." For the associative property, take the first four letters *asso* to stand for "**a**lways **s**tays **s**ame **o**rder" and change the grouping instead.

<u>CO</u>mmutative = <u>C</u>hange <u>O</u>rder.
<u>ASSO</u>ciative = <u>A</u>lways <u>S</u>tays <u>S</u>ame <u>O</u>rder.

Prime and Composite Numbers

A *prime* number is a number that has exactly two factors, 1 and itself. The smallest and only even prime number is 2. Other prime numbers are 3, 5, 7, 11, 13, and so on. A *composite number* is any other whole number greater than 2. A composite number has more than two factors; in other words, it has at least one other number that divides into it without a remainder other than 1 and itself. Examples of these are the numbers skipped in the previous list of prime numbers, such as 4, 6, 8, 9, 10, 12, . . .

The number 1 is considered neither prime nor composite because its only factor is 1. Zero is also considered neither prime nor composite. It is not prime because it is divisible by any number, and it is not composite because it cannot be expressed as the product of two factors that do not include zero.

Factors (GCF) and Multiples (LCM/LCD)

Factors are numbers that divide evenly into a number and do not leave a remainder. For example, the factors of 12 are 1, 2, 3, 4, 6, and 12.

Another related term is the *greatest common factor*, or GCF. This represents the largest number that divides into two or more numbers without leaving a remainder. The greatest common factor of 18 and 24 is 6, because 6 is the largest number that will divide into *both* 18 and 24 without a remainder. The GCF is important when simplifying fractions.

A *multiple* is a positive integer that a whole number divides into without a remainder. For example, multiples of 5 are $5 \times 1 = 5$, $5 \times 2 = 10$, $5 \times 3 = 15$, $5 \times 4 = 20$, . . .

The *least common multiple*, or LCM, of two numbers is the smallest number that each number divides into without a remainder. The LCM of 6 and 8 is 24, because the multiples of 6 are 6, 12, 18, 24, 30, 36,

42, 48 . . . and the multiples of 8 are 8, 16, 24, 32, 40, 48 . . . Looking at the two lists, 24 is the smallest number they have in common.

The least common multiple is also used when adding and subtracting fractions. It is then called the *least common denominator*, or LCD. This value is found the same way as the LCM.

To help keep them straight, remember that factors "fit" into the numbers and multiples are the result of multiplying.

Rational Numbers and Irrational Numbers

The set of real numbers is broken down into two categories, rational and irrational numbers. The *rational numbers* are any numbers that can be written in fraction form. This includes all repeating and terminating decimals, integers, whole numbers, and natural numbers. Remember, any whole number can be written as a fraction by placing the number over 1, as in $4 = \frac{4}{1}$. *Repeating decimals* are decimals where the numbers to the right of the decimal point repeat in the same order indefinitely. These can also be written as fractions, as $0.333333\ldots = \frac{1}{3}$. These numbers can also expressed using a bar over the repeating digits, such as $0.333333\ldots = 0.\overline{3}$.

Irrational numbers are the nonrepeating, nonterminating decimals. These decimals go on forever without stopping or repeating. Some examples are $0.1213856\ldots$, $0.121121112\ldots$, pi (π), and the square root of any nonperfect square. For example, if you change $\sqrt{2}$ to a decimal, the result is $1.414213562\ldots$

Numerator and Denominator

The term *numerator* is the number in the top of the fraction, and the *denominator* is the number in the bottom of the fraction. To keep the terms straight, use the *d* in denominator to remember "down"; the denominator is down below the numerator.

Remember, $\frac{\text{numerator}}{\text{denominator}} = \frac{\text{up}}{\text{down}}$.

Ratio and Rate

Each of these terms compares different things. Specifically, a *ratio* is a comparison of two or more numbers or expressions, such as three girls to four boys. A *rate* is a comparison of two unlike units and often uses the term *per*, such as miles per hour or gallons per minute.

Dividend, Divisor, and Quotient

Each of these terms is used in division problems, and because the first two terms sound so much alike, they are very easily confused. The *dividend* is the value being divided into, the *divisor* is the value being divided by, and the *quotient* is the result of the division or the answer. In the example $35 \div 5 = 7$, 35 is the dividend, 5 is the divisor, and 7 is the quotient.

To remember these terms, it may be helpful to remember the word *divisor* sounds like "divide," and this is the value that is divided by.

Tens Place and Tenths Place

These terms sound alike but have very different meanings. The *tens place* is the place two places to the left of the decimal point. A digit in this place has a value of ten times the digit. The *tenths place* is one place to the right of the decimal point. A digit in this place has a value that is one-tenth the value of the digit.

For example, in the figure provided, a 4 appears in the tens place and a 7 in the tenths place.

Remember that every place value to the right of the decimal point has *ths* at the end of the label. Also, keep in mind that the place values hundreds and hundredths, as well as thousands and thousandths, work in the same way.

Identity and Inverse

Although these mean two different things, you cannot have one without the other. An *identity* element allows you to start with a number or value, perform an operation, and get that same number or value as an answer. In our basic number system, there are two identity elements, and each depends on the operation you are working with. When adding, $5 + 0 = 5$, so 0 is the identity element for addition. When multiplying, $5 \times 1 = 5$, so 1 is the identity element for multiplication.

Think of the *inverse* of a number as the opposite of the number and note that it also depends on the operation you are working with. Remember this: A number and its inverse give the identity as an answer. For example, consider the number 4 and the operation of addition. What number added to 4 would yield the identity of 0? Because $4 + -4 = 0$, -4 is the inverse of 4 when adding.

Take the number 4 again, but change the operation to multiplication. What number would be multiplied by 4 to get the identity of 1? Because $4\times\frac{1}{4}=1$, then $\frac{1}{4}$ is the inverse of 4 when multiplying. Recall that $\frac{1}{4}$ is the reciprocal of 4 because the numerator and denominator are switched.

In summary, 0 is the identity for addition, and 1 is the identity for multiplication. The *additive inverse* of any value is the opposite sign of that value, and the *multiplicative inverse* of any value is the reciprocal of that value.

Square and Square Root

The *square* of a number is the result of that number raised to an exponent of 2. The square of 5 is 5^2, which equals 5×5, or 25. The square root is the opposite. The *square root* of a number is the number that, when multiplied by itself, results in the square of the number. In other words, the square root of 36 is the value that, when multiplied by itself, gives a result of 36. Therefore, because $6 \times 6 = 36$, 6 is the square root of 36.

Radical Sign and Radicand

The *radical sign* $\left(\sqrt{\ }\right)$ is the symbol used to show square roots. For example, the square root of 25 is written as $\sqrt{25}$. This square root is equal to 5. The *radicand* is the value under this symbol. In the example here, the radicand is 25.

Cube and Cube Root

These terms work similarly to the terms *square* and *square root* but use three factors instead of two. The *cube* of a number is the result of that number raised to an exponent of 3. The cube of 4 is $4^3 = 4 \times 4 \times 4 = 64$. The *cube root* is the opposite of this. The cube root of a number is the number that, when multiplied by itself 3 times, results in the cube of the number. In other words, the cube root of 8 is the value that, when multiplied three times, gives a value of 8. Therefore, because $2 \times 2 \times 2 = 8$, 2 is the cube root of 8.

Union and Intersection

These two terms are associated with sets. The *union* of two sets is the set of all elements in both sets put together. The *intersection* of two sets is the set of elements that are common to both sets; in other words, where the sets overlap.

Take the sets $A = \{1, 2, 3, 4\}$ and $B = \{3, 4, 5, 6\}$. The union of these sets, represented by $A \cup B$, is $\{1, 2, 3, 4, 5, 6\}$. These are all of the elements contained in both sets.

The intersection of these sets, represented by $A \cap B$, is $\{3, 4\}$. These are the only two elements that are common to both sets.

Think of a family reunion when trying to remember union. Everyone from *all* sides of the family is included and invited to the reunion. Note that the symbol for union looks like a *u*. To remember intersection, think of the intersection of two roads as the only place where the two roads overlap, the part the roads have in common. Thus, the intersection of two sets includes only the values in both sets.

Algebra

Constant and Variable

A *constant* is anything that will not change in value, so any number represents a constant in math. For example, 5 is a constant, as well as -6.7. A *variable* is anything that could change, or vary. Variables in math are often represented as letters, such as x and y, but could also be represented by other symbols, such as circles and squares.

Expression and Equation

Each of these terms contains numbers, symbols, and operations. The difference between them is that *equations* contain an equal sign and *expressions* do not. A way to view an equation is as two expressions set equal to each other. The phrase $3x^2 - 4x + 2$ is an expression. The sentence $4x + 8 = 5x$ is an equation.

Equation and Inequality

An *equation* is a math sentence that shows that two expressions are equal. An equation always contains an equal sign. An *inequality* shows that two expressions are not equal. This can be shown by using one of five different signs instead of an equal sign: greater than ($>$), greater than or equal to ($\geq$), less than ($<$), less than or equal to ($\leq$), or not equal ($\neq$). For example, $3x + 4 = 12$ is an equation, while $4x + 6 \leq 10$ and $2x > 10$ are inequalities.

Simplify and Solve

These two terms confuse many students because they think they mean the same thing. The word *simplify* means to make smaller and easier to use. For example, you could simplify an expression by multiplying and combining like terms, but you may still not know the value of the variables in the problem. The term *solve* specifically means that an unknown will be found at end of the problem. Equations can be solved for the value of a specific letter, or variable, because they contain an equal sign. Expressions are simplified; because they do not have an equal sign, they cannot be solved.

Cross-Multiplying and Cross-Canceling

Cross-multiplying is a process used when solving a proportion where the numerator of the first fraction is multiplied by the denominator of the second fraction. This is then set equal to the result of multiplying the denominator of the first fraction by the numerator of the second fraction. The result is an equation that can be solved.

In the equation provided here, the proportion can be cross-multiplied to get $2x = 20$. Then, it can be divided by 2 to get $x = 10$.

$$\frac{x}{4} = \frac{5}{2}$$

$$2x = 20$$
$$x = 10$$

Cross-canceling is a strategy used when multiplying fractions. In this process, common factors are divided, or canceled out of the numerators and denominators. In the following problem, the numerator of one fraction and the denominator of the other fraction have a common factor of 2. By dividing out this common factor and then multiplying the fractions together, the final answer is in simplest form.

$$\frac{4}{7}\times\frac{3}{8}=\frac{\cancel{4}^{1}}{7}\times\frac{3}{\cancel{8}_{2}}=\frac{3}{14}$$

Linear and Quadratic Equations

Equations that are *linear* contain a variable where the greatest exponent is 1. In other words, there are x's but no x^2, x^3, or other larger exponents in the equation. The equation $y = 2x + 1$ is a linear equation. When linear equations are put on a graph, the result is always a straight line. An example is shown here.

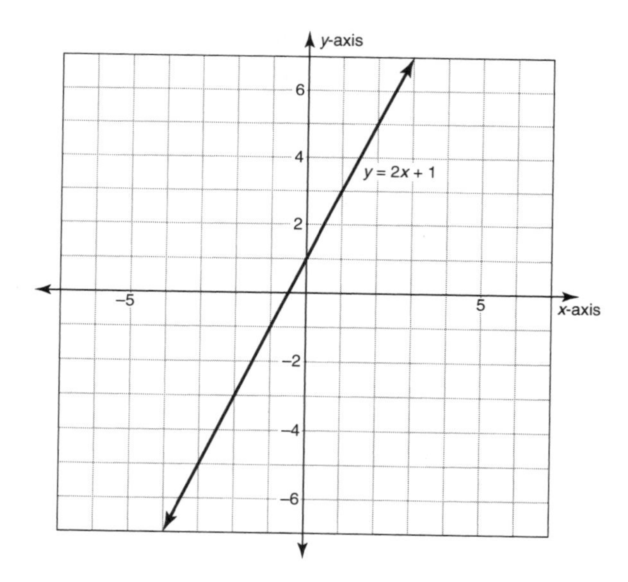

A *quadratic* equation has a highest exponent of 2 on its variables. The equation $y = 3x^2 + 4x - 6$ is a quadratic equation. When quadratic equations are graphed, the result is a U-shaped graph. An example of a quadratic equation is graphed here.

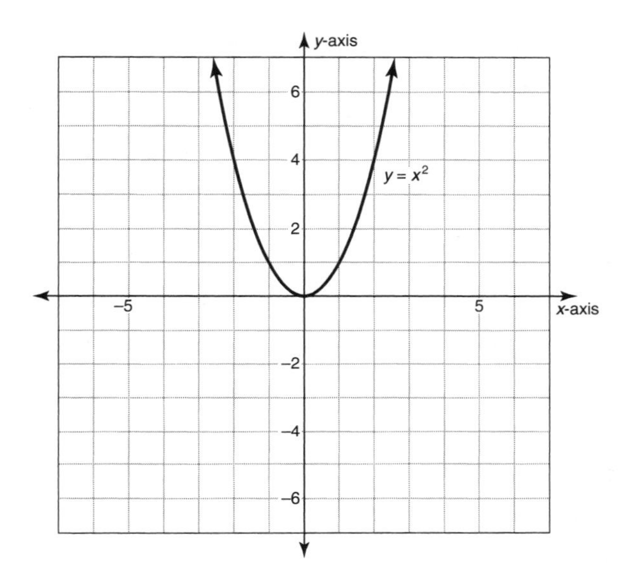

Geometry and Measurement

Complementary and Supplementary Angle Pairs

Each of these terms describes a pair of angles. A *complementary* pair is two angles whose sum is 90 degrees. If this pair of angles is adjacent, or next to, each other, then the angles together form a right angle in the shape of an *L*. The following diagram shows a pair of adjacent and a pair of nonadjacent complementary angles.

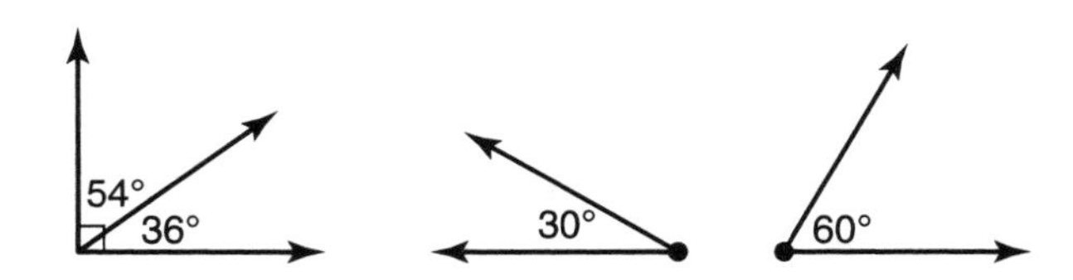

A *supplementary* pair is two angles whose sum is 180 degrees. If the angles in this pair are adjacent to each other, the angles form a straight line and are also referred to as a *linear pair*. The diagram here shows a pair of adjacent and a pair of nonadjacent supplementary angles.

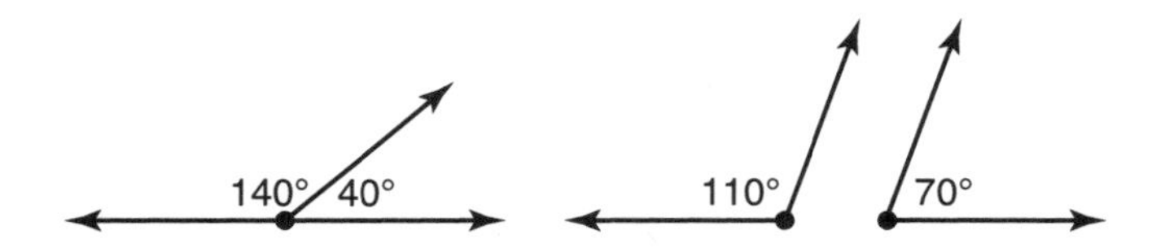

One way to remember the difference is to think that the *c* in complementary comes before the *s* in supplementary in the alphabet, and 90 comes before 180 in the number system. Another way is to make a 9 out of the *c* in complementary by drawing a vertical line and adding a 0 to make 90. Then, make an 8 out of the *s* in supplementary and form the number 180.

Congruent and Similar

Two figures that are *congruent* are exactly the same in every way. When visualizing congruent shapes, think of figures that would appear to be twins. *Similar* figures are the same shape and proportion but are different sizes. For example, two triangles that are similar would look the same, but the length of the sides would be 2 or 3 times bigger in the larger triangle.

In the figure shown here, triangles 1 and 2 are congruent, and triangle 3 is similar to triangle 1 and triangle 2.

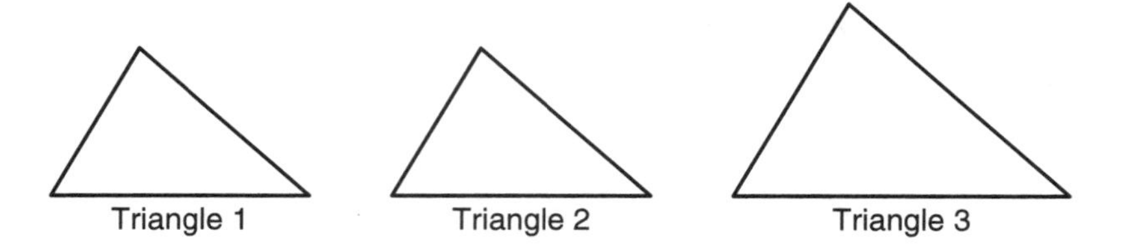

Line and Line Segment

A *line* is a straight path that continues in both directions forever. A *line segment* is part of a line; it is two points on a line and the points between them. The length of a line segment can be measured, but the length of a line cannot because it continues an infinite distance.

Line AB is written $\overleftrightarrow{AB}$.
Line segment AB is written $\overline{AB}$.

Parallel and Perpendicular Lines

Parallel lines are two lines in the same plane that will never intersect, or cross. *Perpendicular* lines intersect at right angles. When thinking of parallel lines, use real-world examples such as train rails or the double yellow line on many roads. The lines that cross to form a checkerboard are a good example of perpendicular lines. An example of each is shown here.

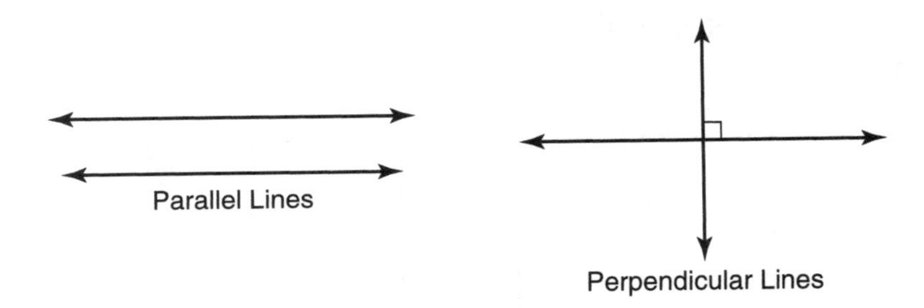

Units and Square Units

A *unit* is a linear measure; it is a part of a line used to measure the length of something. If you were to measure the length of a room, the height of a person, or the perimeter of a garden, you would use regular units.

Square units are two-dimensional units used to find the area of a region. Square units are actually made up of squares; the number of squares that would fit in a certain region or shape determines the area of the region. If you were trying to find the amount of paint needed to cover a wall, you would use square units.

Perimeter and Area

Perimeter is the distance around an object, or the sum of its sides. The formula for perimeter is commonly known as P = side + side + side + . . . If you were to take a walk around the outside of a playground, you would calculate the perimeter to figure out how far you walked. Perimeter is measured in linear units.

The *area* of a figure is the number of squares that would cover a particular region. The area is calculated by multiplying the length times the width of the figure and is measured in square units. When calculating the amount of carpet needed for a room, you would find the area of the floor of the room in square units.

The following figure illustrates the difference between perimeter and area.

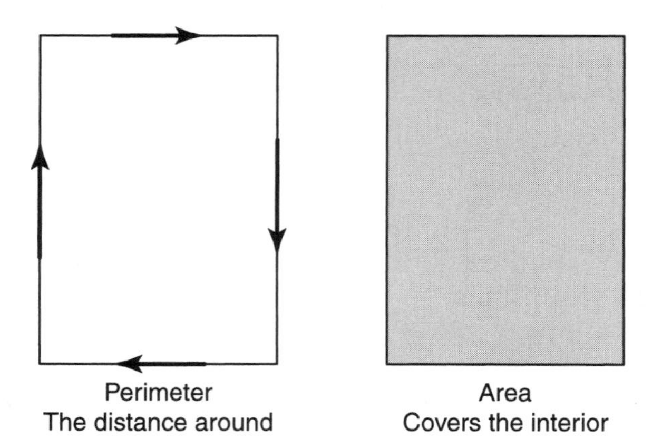

Diameter and Radius

These are two terms used when working with circles. The *diameter* is a line segment, or *chord*, located in the circle that goes through the center of the circle and touches the circle in two places. The diameter cuts the circle into two equal parts, half-circles called *semicircles*. The *radius* of a circle is half of the diameter. A radius is a line segment that connects the circle with its center. In the following diagram, line segment $\overline{AO}$ is a radius, and line segment $\overline{AB}$ is a diameter.

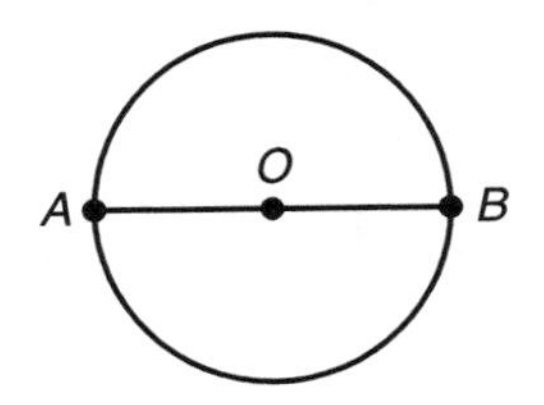

Circumference and Area of Circles

Think of the *circumference* of a circle as the perimeter of the circle. Because a circle does not have sides, this term is used to describe the distance around. The formula for circumference is $C = \pi \times$ diameter (πd) or $C = 2\pi \times$ radius ($2\pi r$).

The *area of a circle* is the number of square units needed to cover the circle. Because squares are used to fill a circular region, rounding is used when calculating the area. The formula for the area of a circle is $A = \pi \times$ radius $\times$ radius (πr^2).

These formulas are easily mixed up because they are so similar. To keep them straight, remember the saying "**C**herry **pie** is **d**elicious" ($C = \pi d$) for the **c**ircumference formula, and "**A**pple **pie**s a**r**e **too** (two)" ($A = \pi r^2$) for the formula for **a**rea.

Surface Area and Volume

The *surface area* of a three-dimensional figure is exactly what it says: the surface located on the outside of the figure. To help with the concept of area, think of the amount of wrapping paper needed to cover a box. The *volume* of a three-dimensional figure is the amount of space inside the figure. When visualizing volume, think of the amount of material that would fill a box.

x-axis and *y*-axis

The *x-axis* is the horizontal number line on a graph, and the *y-axis* is the vertical number line on a graph. Each is shown on the following diagram.

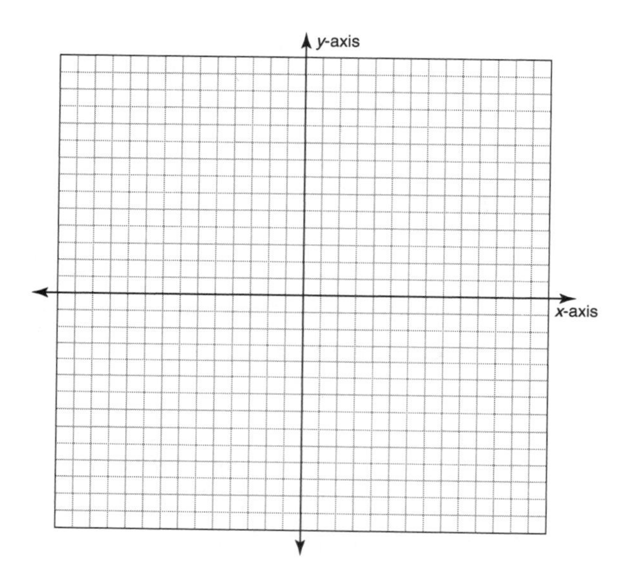

x-coordinate and *y*-coordinate

Points, in coordinate geometry, are always written as a pair of numbers surrounded by parentheses. The *x-coordinate* is the first value in a coordinate pair. This value, which is also known as the *abscissa*, tells the number of units that are counted along the *x*-axis to the right or left of zero when graphing a point. The *y-coordinate* is the second value in a coordinate pair. This value, which is also known as the *ordinate*, tells the number of units that are counted up or down from zero along the *y*-axis from zero when graphing a point. Positive values are counted to the right and up; negative values are counted to the left and down.

To help remember this order, think that (x, y) is in alphabetical order. In the following figure, the point (3, 1) is graphed by starting at the origin and counting 3 units to the right and 1 unit up. The point (1, 3) is graphed by starting at the origin and counting 1 unit to the right and 3 units up.

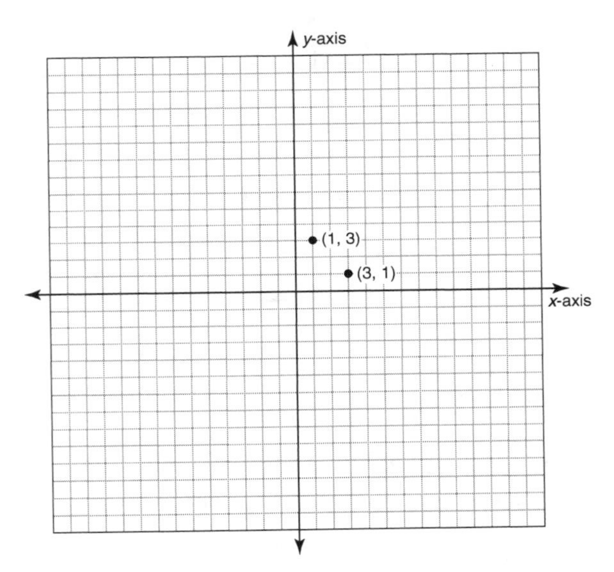

Transformations and Translation

A *transformation* in geometry is a general term used to describe different changes that are made on a figure. Many types of transformations are studied in geometry. One of these transformations is known as a *translation.* A translation takes a figure and "slides" it to a new location in the plane or coordinate grid. For example, triangle ABC in the figure provided here is translated 3 units to the right and 2 units down to become triangle $A'B'C'$.

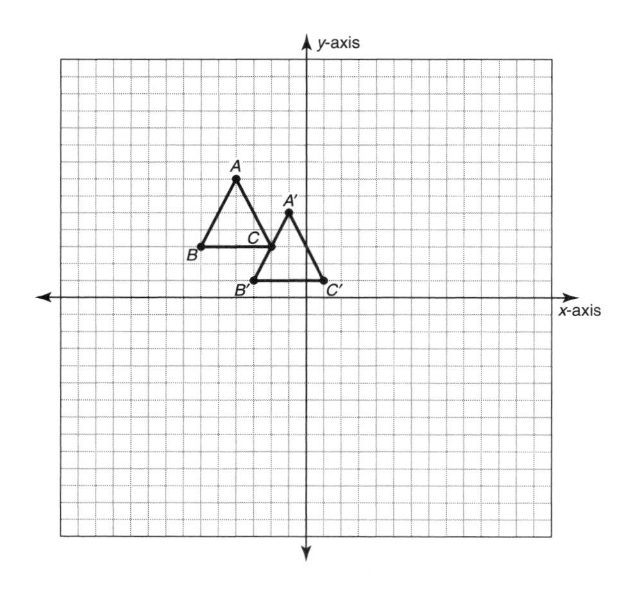

Hypotenuse and Leg

These two terms represent parts of a right triangle. The *hypotenuse* is the longest side of a right triangle and will always be the side across from the right triangle. There are two *legs* of any right triangle; these are the sides that form the right angle of the triangle. The following diagram shows the placement of each in triangle *ABC.*

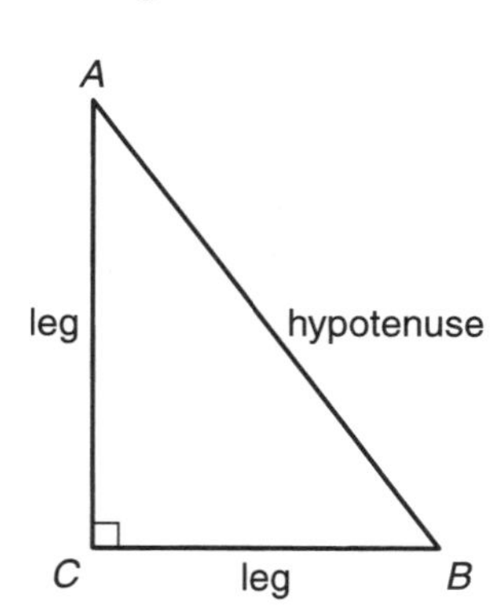

Probability and Statistics

Permutations and Combinations

Each of these calculations is figured out in a similar fashion, and they are easy to confuse. Remembering what each of them means can be a big help in calculating them correctly. A *permutation* is the number of different arrangements of a group of objects; every time the order of the objects changes, this creates a different permutation of the objects. For a group of three students arranged from a group of five students, there are $5 \times 4 \times 3 = 60$ different orders or permutations of these students.

A *combination* is the number of ways part of a set can be grouped; in a combination, the order does not matter. Because the order does not change the combination of items, you need to divide by the different ways the items can be arranged. A combination of three students chosen from a group of five is equal to $\frac{5\times4\times3}{3\times2\times1} = \frac{60}{6} = 10$ ways.

For example, selecting three class officers from a group of five students is a permutation. Let's say the first person selected is the president, the second person selected is the vice president, and the third person selected is the treasurer. Therefore, the order that the students are selected determines the office they hold; order is important. One permutation might have a certain student as president, and a different permutation would have that same student as treasurer. Every time the order changes, a different permutation is made. As shown previously, there are 60 ways to do this.

An example of a combination would be the number of ways three students can be selected from a group of five to be members of a school club. In this case, the order is not important. The number of combinations would not change if the order of the students selected changed, and it would not matter who was selected first, second, or third. As shown in the previous example, there are only 10 combinations of three students chosen from five.

Mean, Median, and Mode

These terms represent different calculations that give information about a set of data. They are called *measures of central tendency*. The *mean* is found by adding each of the values in a data set and then dividing by the total number of values.

The *median* is the middle value when the set is placed in order. If two numbers share the middle spot in the list, then the mean of those two values becomes the median. This occurs anytime the list has an even number of values.

The *mode* is the value that occurs, or repeats, most often in the set.

To help remember this, use the fact that the term *median* and the word *middle* each have the same number of letters to help relate them. Median can also be remembered as the median strip that runs down the middle of a highway.

To remember mode, think about how the first part of the term *mode* sounds similar to the word *most*.

How Can You Help Your Child Learn Math Vocabulary?

Learning correct math vocabulary is the first step in mastering math skills and understanding the concepts. Because the vocabulary terms are so precise, you and your child need to lay a strong foundation to ensure success.

Here are some strategies to help your child both learn new vocabulary and reinforce the meanings of words she has already encountered. These strategies will provide multiple approaches so that your child does not get discouraged or flustered when encountering an unfamiliar word or concept.

Help your child activate prior knowledge of the term. For example, if a certain topic was taught in a prior grade or even earlier in the school year, try to help the student remember where and when a concept was studied. Discuss the types of problems he encountered and the ways that these particular types of problems were solved. Your child's teacher can be a big help in figuring out when and where certain concepts were taught.

Use context clues to assist in finding the meaning of a word. When given a word problem, the student should look for key words and hints as to the process needed to solve the problem. Chapter 4 discusses these hints and key phrases in detail.

Talk to your child about the structure of words. Many terms in mathematics have prefixes, suffixes, or roots that help define the meaning of the word. Show him or her, for example, how the term *triangle* is made up of the prefix *tri* and the root word *angle*. The prefix *tri* means "three," and a triangle is a shape with three angles. Connect this idea with a common fact, such as a *tricycle* has three wheels, to help reinforce the definition.

Use a notebook or binder and start a math vocabulary dictionary that the student can keep at home or use at school as a reference. Include in the notebook the main terms that the student encounters during a homework assignment, or a certain unit, especially those words she has a difficult time with. For each term, have your child first list the book definition. Then, have the student rephrase the definition in her own words; that is, how she would explain the meaning to a friend, sibling, or even you. The student should also include a diagram or drawing of the term along with the definition because math is such a visual subject.

Write the words from the math vocabulary dictionary on index cards, putting the word or term on one side and the definition on the other. Use these flashcards to practice the terms on a regular basis.

If the student is issued a textbook, show him where the index and glossary are located in the text. Demonstrate how to look up words in the index to find places in the book where the topic is discussed and problems are solved. A good math glossary will provide definitions, as well as examples and pictures, to help your child attain and retain the meaning of the word.

Show your child how to reread a question to help with understanding. Students can also learn to rephrase a question to help a word make sense.

A word of caution here is to minimize the number of new words that a student is learning at one time. As presented earlier in this chapter, some of the terms, especially the synonyms, can be confusing, and your child will have more success if the concepts are studied in smaller "chunks."

Establishing good habits and a positive attitude when learning vocabulary is important, and it will help lay the foundation for strong math skills. Confidence in this area will help your child get over the hurdles of word problems and complicated scenarios that are common when solving math questions.

Chapter 3

The Rules You Should Know

There are a number of basic math rules that you should be familiar with so that you can recognize if your child is breaking any of them. The most basic operations are addition, subtraction, multiplication, and division, and most of the other rules have elements of these operations embedded in them.

The basics in this chapter form the foundation for much of the future concepts your child will encounter through math studies. Each rule in this chapter is followed by a current example of its use. In addition, an example from high school or entry-level college applications is included. This exposure will help you and your child see the importance of mastering these concepts and illustrate the interconnectedness of math study.

Order of Operations

The *order of operations* is an agreed-upon order of how to perform the basic math operations so that answers are consistent. It is helpful first to illustrate why a dictated order of operations is needed. For example, in the problem $6 + 10 \div 2$, two different results could be obtained, depending on which operation you perform first. If you incorrectly perform addition first, then $6 + 10 \div 2 = 16 \div 2 = 8$. If you correctly perform division first, you get the correct answer of $6 + 10 \div 2 = 6 + 5 = 11$.

You may recall from your school math experiences that the order of operations can be remembered by the acronym PEMDAS, or "Please excuse my dear Aunt Sally." The order is as follows:

- P — **P**arentheses and grouping symbols
- E — **E**xponents
- MD — **M**ultiplication and **D**ivision from left to right
- AS — **A**ddition and **S**ubtraction from left to right

Emphasize to your child the fact that parentheses can indicate many grouping symbols, such as ( ), { }, [ ], and the fraction bar. The square root symbol, √, is really an exponent and is evaluated at that level. Here are some examples of the order of operations:

Example 1: $20 - 5 + 4 =$

$15 + 4 =$

19

In this example, evaluate subtraction first, because addition and subtraction are done from left to right.

Example 2: **$20 - (5 + 4) =$**
$20 - 9 =$
11

In example 2, first evaluate the parentheses and then subtract.

Example 3: **$100 + 20 \times 3 =$**
$100 + 60 =$
160

Multiplication is evaluated before addition.

The fraction bar serves as a grouping symbol, so when simplifying a fraction, evaluate the numerator, evaluate the denominator, and then finally divide. For example, a fraction expression such as $\frac{5+12}{3-1}$ is interpreted as $\frac{(5+12)}{(3-1)} = \frac{17}{2}$.

Looking Ahead

In high school algebra, your child will have to use the order of operations to solve problems such as converting a temperature from Fahrenheit to Celsius using the formula $C = \frac{5}{9}(F-32)$. In this formula, first subtract 32 from the Fahrenheit temperature, because of the parentheses, and then multiply the result by $\frac{5}{9}$. Another complex example is the formula for the roots of a quadratic equation: $x = \frac{-b \pm \sqrt{b^2 - 4a}}{2a}$. In this formula, first evaluate the numerator because a fraction bar is a grouping symbol. The numerator has a square root sign (grouping symbols), so that is evaluated first. (Square roots were defined in the vocabulary list from chapter 2.) Within the square root, square the value of b, (exponents); multiply 4, a, and c (multiplication); and then find the difference. Still working on the numerator, take the opposite of b. (Integer arithmetic is defined later in this chapter.) In the denominator, multiply two times the value of a.

Finally, divide the numerator by the denominator to get the resultant roots of a quadratic equation. You can see by this example how important it is to master the order of operations!

Exponents and Exponent Rules

Exponents are a math concept that involves repeated multiplication of the same factor. The concept of exponents is introduced in the upper elementary grades, and then the rules for working with exponents are studied in middle school. These rules are used in advanced math study.

A *power* is a numerical or variable expression that contains a base factor and an exponent. For example, the numerical expression 2^3 is the third power of 2, and $2^3 = 2 \times 2 \times 2 = 8$. Therefore, 8 is considered the third power of 2. In this expression, 2 is the base and 3 is the exponent. To perform operations on powers, there are rules that dictate how you calculate.

Multiplying Powers with the Same Base Factor

When multiplying powers with the same base, keep the base factor and add the exponents.

For example, look at $4^3 \times 4^5 = 4^{3+5} = 4^8$. In another example, $x^2 \cdot x^7 = x^{3+5} = x^8$, or, for any real numbers a and b, $x^a \times x^b = x^{a+b}$.

Expanding the exponents illustrates this better. Take the previous power example $4^3 \times 4^5$. By expanding, this expression becomes

$4^3 \times 4^5 = (4 \times 4 \times 4) \times (4 \times 4 \times 4 \times 4 \times 4)$. You can easily see that there are now eight factors of 4, which is 4^8.

Dividing Powers with the Same Base Factor

When dividing powers with the same base, keep the base factor and subtract the exponents.

Let's look at an example with division of powers, expand out the exponents, and determine the rule for dividing. Consider the fractional expression $\frac{8^6}{8^4} = \frac{8\times8\times8\times8\times8\times8}{8\times8\times8\times8}$. Use canceling and cancel out four of the eights in the numerator with four of the eights in the denominator. You are left with 2 eight factors in the numerator and the value of 1 in the denominator. $\frac{8^6}{8^4} = \frac{8\times8\times8\times8\times8\times8}{8\times8\times8\times8} = \frac{8\times8}{1} = 8^2$. Through the process of canceling common factors, you have in essence subtracted the value of the exponent in the denominator from the value of the exponent in the numerator. In another example, $\frac{n^{12}}{n^5} = n^{12-5} = n^7$, or, for any real numbers a and b, $x^a \div x^b = x^{a-b}$.

The Power of a Power

When raising a power to another power, keep the base and multiply the exponents.

For example, $(2^3)^2$ means that there are two groups of 2 to the power of 3: $2^3 \times 2^3 = (2 \times 2 \times 2) \times (2 \times 2 \times 2) = 2^{3\times2} = 2^6$. In another example, $(m^4)^3 = m^{12}$, or, for any real numbers a and b, $(x^a)^b = x^{a\times b}$.

The Power of a Product

When raising a product within grouping symbols to a power, the exponent applies to every factor in the group.

For example, $(5x)^3 = 5^3 \times x^3 = 125x^3$. For any real number a, $(mn)^a = m^a n^a$.

The Power of a Quotient

When raising a quotient within grouping symbols to a power, the exponents applies to both the dividend (numerator) and the divisor (denominator) of the quotient.

For example, $\left(\frac{x}{4}\right)^3 = \frac{x^3}{4^3} = \frac{x^3}{64}$.

Looking Ahead

When studying algebra, students are expected to use a combination of the properties covered in this chapter in the same question. For example, to multiply the two fractions $\frac{(3x)^2}{2x} \times \frac{4x^5}{x^3}$, begin by applying the exponent in the first numerator. The power of a product property coverts the expression $(3x)^2$ to $9x^2$. Then, multiply the numerators and the denominators across as in any fraction. Remember to use the rule for multiplying with exponents with the same base, which says to add the exponents. Multiply and divide the whole numbers just as in any problem. The expression now becomes $\frac{9x^2}{2x} \times \frac{4x^5}{x^3} = \frac{36x^7}{2x^4}$. Finally, divide the whole numbers, use the exponent rule for dividing with like bases. The simplified expression is $18x^3$.

Properties of Numbers

In addition to the order of operations, your child needs to understand some basic properties of numbers that allow the order of operations to be changed. These properties include the associative property, the commutative property, and the distributive property.

The Associative Property

The order of operations instructs you to evaluate expressions in parentheses, or grouping symbols, first. The *associative property* of addition or multiplication allows you to change the grouping without changing the resultant sum or product. For example, if a problem states $(9 + 11) + 4$, the order of operations instructs you first to add 9 and 11, which equals 20, and then add 4 for a result of 24. The associative property allows you to change the grouping to be $9 + (11 + 4)$. By using this property, you would now first add 11 and 4, which equals 15, and then add 9 for the same result of 24. Similarly, you can use the associative property when multiplying. Consider this problem: $5 \times (2 \times 18)$. Order of operations directs you to multiply 2 and 18 first. However, the arithmetic would be simpler if you first multiplied 5 and 2 to get 10. The associative property assures you that you can change this grouping and still arrive at the correct answer:

$5 \times (2 \times 18)$	**The original expression**
$(5 \times 2) \times 18$	**The associative property**
10×18	**Evaluate in the grouping symbols**
180	**Answer arrived at using easier computation**

The synonym section of chapter 2 also covers this property because it is often confused with the commutative property. The associative property can be remembered by the first four letters:

ASSOciative = **A**lways **S**tays **S**ame **O**rder.

The Commutative Property

According to the order of operations, add or subtract terms in an expression from left to right. However, the *commutative property* of addition or multiplication allows you to change the order of addends in an addition problem or change the order of factors in a multiplication problem. For example, in the problem 16 + 37 + 14, you would first add 16 + 37 by following the order of operations. Notice, however, that 16 + 14 = 30, so if you change the order of the addends, the addition becomes easier to evaluate. The commutative property enables you to do this:

16 + 37 + 14	**The original expression**
16 + 14 + 37	**The commutative property**
30 + 37	**Evaluate from left to right**
67	**Answer arrived at using easier computation**

Likewise, you can change the order in a multiplication problem. For example, in the expression 20 × 19 × 5, you can make the computation easier if you recognize that 20 × 5 = 100. Reorder the expression to be 20 × 5 × 19 = 100 × 19 = 1,900.

This property is also covered in the synonym section of chapter 2 because it is often confused with the associative property. The commutative property can be remembered by the first two letters:

COmmutative = **C**hange **O**rder.

The Distributive Property

In a problem such as 12 (5 + 10), the order of operations directs you first to add the terms in parentheses, then to multiply by 12. However, the *distributive property* allows you to multiply each term in parentheses by 12 first and then add as the final operation.

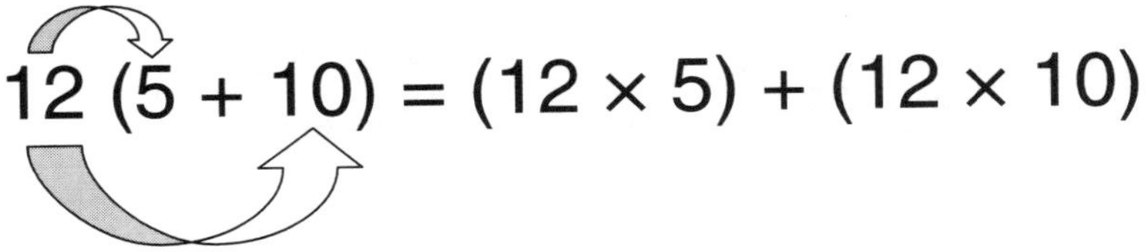

$$12\ (5 + 10) = (12 \times 5) + (12 \times 10)$$

Therefore, you can verify that 12 (5 + 10) = 12 (15) = 180 gives the same result as 12 (5 + 10) = (12 × 5) + (12 × 10) = 60 + 120 = 180. This property distributes the factor of 12 to every term in parentheses. In later study, the distributive property will be a necessary rule to follow to manipulate variables in equations.

You can help your child to remember this property with the word *distribute*. When a teacher distributes worksheets, he gives one to each person in the room. In much the same way, the factor before the parentheses in an expression is "distributed," or multiplied, to every term within the parentheses.

Looking Ahead

The properties of numbers are integral parts of algebra and solving equations, which your child will encounter in high school. For example, to solve the equation $7(x - 4) = 3x$, you cannot combine the x and the -4 terms, so you must use the distributive property to solve the equation.

Furthermore, you must use the commutative property to combine like terms:

$7(x-4)=3x$	
$7x-28=3x$	**Use the distributive property.**
$7x-28-3x=3x-3x$	**Subtract $3x$ from both sides.**
$7x-3x-28=3x-3x$	**Use the commutative property.**
$4x-28=0$	**Combine like terms.**
$4x-28+28=0+28$	**Add 28 to both sides.**
$\frac{4x}{4}=\frac{28}{4}$	**Divide both sides by 4.**
$x=7$	**The solved equation**

Solving equations is described later in this chapter.

Fraction Arithmetic

Fractions are one of the most challenging concepts for students and one of the biggest sources of math anxiety. Often, students immediately "turn off" when they see a fraction in a math problem. A good basis in understanding fractional arithmetic is crucial for your child's confidence in math.

Adding and Subtracting Fractions

Adding and subtracting fractions requires a common denominator. The best common denominator is the *least common multiple* (LCM) of the denominators, which is called the *least common denominator* (LCD). Find the LCD of the given denominators, and make equivalent fractions with this denominator. Then add or subtract the numerators and keep the denominator. Finally, express the fraction in its simplest form, where there are no common factors in the numerator and denominator. For example, in the problem $\frac{1}{2}+\frac{3}{5}$, the LCD is 10. Convert each fraction to have a denominator of 10: $\frac{1}{2}\times\frac{5}{5}=\frac{5}{10}$, and $\frac{3}{5}\times\frac{2}{2}=\frac{6}{10}$. The problem is now $\frac{5}{10}+\frac{6}{10}$. Add the numerators and keep the denominator: $\frac{5+6}{10}=\frac{11}{10}$, or the equivalent mixed number $1\frac{1}{10}$.

Subtraction works the same way. For example, solve $\frac{13}{15}-\frac{3}{20}$. Begin by finding the LCD of 15 and 20, which is 60. Convert each fraction to have a denominator of 60: $\frac{13}{15}\times\frac{4}{4}=\frac{52}{60}$, and $\frac{3}{20}\times\frac{3}{3}=\frac{9}{60}$. Then rewrite the problem as $\frac{52}{60}-\frac{9}{60}$. Subtract the numerators and keep the denominator: $\frac{52-9}{60}=\frac{43}{60}$.

Multiplying Fractions

Multiplying fractions is actually easier than adding or subtracting. Simply multiply the numerators straight across and multiply the denominators straight across as well. Then simplify the fraction, if needed. For example, $\frac{2}{3}\times\frac{5}{7}=\frac{2\times5}{3\times7}=\frac{10}{21}$. Sometimes, multiplication can be made easier by first applying the concept of canceling. *Canceling* is just the process of putting the product in lowest terms prior to multiplying. Find a common factor to any numerator and any denominator and divide this factor out before multiplying. For example, look at this problem: $\frac{12}{45}\times\frac{15}{28}$. The numbers 15 and 45 have a common factor of 15, and the numbers 12 and 28 have a common factor of 4. Cancel out these common factors as shown: $\frac{\cancel{12}^{3}}{\cancel{45}_{3}}\times\frac{\cancel{15}^{1}}{\cancel{28}_{7}}=\frac{3\times1}{3\times7}=\frac{3}{21}=\frac{1}{7}$.

Understanding the helpful method of canceling makes multiplication easier. It is also a necessary skill to master for later work in algebra. Help your child avoid common traps, such as trying to cancel a common factor between two numerators. This will not result in a correct answer.

Another pitfall is trying to cancel when there are addition or subtraction operations in the numerator or denominator. For example, in the expression $\frac{12+7}{6}$, you cannot cancel out a common factor of 6 because of the addition operator in the numerator. The addition must be performed first by using the order of operations because a fraction bar is a grouping symbol.

Dividing Fractions

Sometimes, division of fractions can be accomplished as easily as multiplication by dividing straight across: $\frac{10}{45} \div \frac{2}{5} = \frac{5}{9}$. Most often, however, the numbers are not so agreeable, and straight division would result in remainders. In these cases, take advantage of the concept that division is the opposite operation, or inverse function, of multiplication. Therefore, doing two opposite operations will yield the same result as dividing. To divide fractions, change the division sign to a multiplication sign and multiply by the reciprocal ("the flip") of the divisor (the second fraction). For example, $\frac{8}{15} \div \frac{2}{7} = \frac{8}{15} \times \frac{7}{2} = \frac{56}{30} = \frac{28}{15}$.

Looking Ahead

In high school, your child will face algebraic manipulation of fractions, which uses the same basic concept but with variables in the numerator and denominator. For example, to add $\frac{5}{x} + \frac{3}{x^2}$, find the common denominator of x^2 and find an equivalent fraction: $\frac{5}{x} \times \frac{x}{x} = \frac{5x}{x^2}$. Now, add the fractions to get $\frac{5x}{x^2} + \frac{3}{x^2} = \frac{5x+3}{x^2}$.

Fractional arithmetic is also used in advanced math study of probability, where you must add or multiply the probability ratios.

For example, to find the probability of flipping a coin that lands heads, and then rolling a six-sided die to get a five, you would multiply $\frac{1}{2}\times\frac{1}{6}=\frac{1}{12}$.

For fractional arithmetic, remember the following:

- Adding and subtracting fractions requires a common denominator.
- Multiplying and dividing fractions can be simplified by canceling common factors.

Using Cross-Multiplication with Ratios and Proportions

A *ratio* is a numerical comparison of information; for example, the number of girls to the number of boys in a classroom or the number of girls to the total number of students. Ratios can be written as a fraction, $\frac{\text{girls}}{\text{total}}$; with the words *girls to total*; or with a colon, girls:total. Ratios can be simplified just as fractions are. A *proportion* is an equation that states that two ratios are equal. For example, if there are three girls to every two boys in the school, and there are 300 girls in the school, then there are 200 boys because the proportion $\frac{3}{2}=\frac{300}{200}$. In a proportion, it is said that the product of the means is equal to the product of the extremes.

$$\frac{3}{2}=\frac{300}{200}$$

In the proportion shown in the figure, you can verify that $3 \times 200 = 2 \times 300$. This is also referred to as cross-multiplication. You can use this fact about proportions to solve many problems involving ratios. To solve for an unknown quantity involving ratios, set up a proportion and cross-multiply. For example, find the value of n in a proportion:

$\frac{24}{64} = \frac{n}{10}$	**The given proportion**
$24 \times 10 = 64n$	**Cross-multiply.**
$240 = 64n$	**Multiply.**
$\frac{240}{64} = \frac{64n}{64}$	**Divide both sides by 64.**
$n = 3.75$	**Isolate the variable.**

Looking Ahead

Proportions are used to help solve many different types of problems in math, including geometry, advanced algebra, and percent problems. Consider the following example:

Two similar triangles are in the ratio of 4:5. If the shortest side of the larger triangle is two more than the shortest side of the smaller triangle, what is the length of the smallest side of each triangle?

One approach to solving this problem is to let x equal the length of the shortest side of the smaller triangle and let $x + 2$ equal the length of the shortest side of the larger triangle. Then, set up the proportion that relates these two sides to the ratio of the two triangles.

$\frac{x}{x+2} = \frac{4}{5}$	**The proportion that relates the sides to the given ratio**
$5x = 4x + 8$	**Cross-multiply and use the distributive property.**
$5x - 4x = 4x - 4x + 8$	**Subtract 4x from each side.**
$x = 8$	**Simplify.**

The shortest side of the smaller triangle is 8 units, and the shortest side of the larger triangle is $x + 2 = 8 + 2 = 10$ units.

Percents

A *percent* is a way to write a number, usually less than 1, as a ratio comparing a value to 100. After all, *per* means "for each" and *cent* means "100."

Fractions and Percents

To change a fraction to a percent, solve the proportion: $\frac{\text{part}}{\text{whole}} = \frac{\%}{100}$. For example, $\frac{2}{5}$ is 40%, because $\frac{2}{5} = \frac{n}{100}$. Cross-multiply to get $200 = 5n$, divide both sides by 5, and $n = 40\%$. To change a percent to a fraction, express the percent as a fraction by placing the percent over 100 and then find the simplest form of the fraction. For example, 56% is $\frac{56}{100} = \frac{14}{25}$.

Decimals and Percents

Because the decimal system is a place value based on powers of 10, it is a simple matter to change a decimal value to a percent. Just multiply the decimal by 100, which is the same as moving the decimal point two places to the right. For example, 0.67 is 67%. Likewise, to change from a percent to a decimal, just divide by 100, which is the same as moving the decimal point two places to the left. Therefore, 36.4% is the number 0.364.

Looking Ahead

Percent concepts arise in later math studies when word problems involve percent increases and decreases, such as sales and tax applications. For example, to find the discount for a $35 pair of pants on sale at 40% off, change the 40% to the decimal equivalent 0.40 and multiply this by $35 to get $0.40 \times 35 = \$14$ discount. Percents are also used to create circle graphs, which show parts of a whole with statistical data.

Integer Arithmetic

Integers are the whole numbers and their opposites {. . . –4, –3, –2, –1, 0, 1, 2, 3, 4 . . .}. It is important to understand the rules for performing the basic operations of addition, subtraction, multiplication, and division with integers. An integer has a magnitude and a sign. Think of the *magnitude*, also called the *absolute value*, as the distance that the integer is from zero on a number line. Therefore, the absolute value of an integer, represented as $|x|$, is the numerical value of the integer and is always positive. Think of the *sign* as the direction from zero on a number line. If the number is to the right of zero, the sign is positive, and if the number is to the left of zero, the sign is negative.

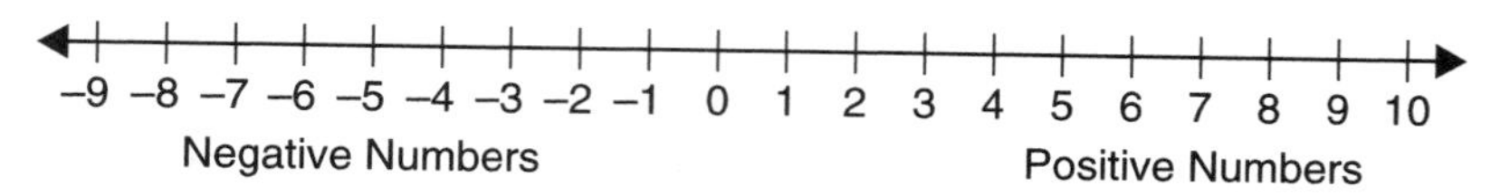

Adding Integers

Consider a number line to help clarify the rules for adding integers. To add 5 + 3, start at zero on the number line and travel 5 units to the right because it is positive. From there, travel 3 more units to the right to end at +8. When you add two positive integers, add the absolute values and keep the result positive, as shown in the following example:

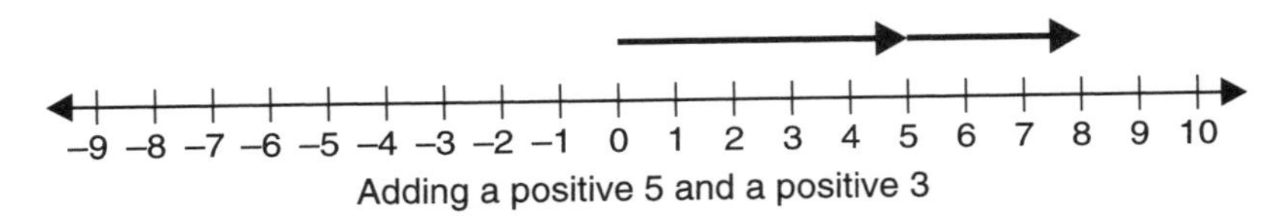

Adding a positive 5 and a positive 3

To add –2 + –4, start at zero on the number line, and travel 2 units to the left because it is negative. From there, travel 4 more units to the left to end at –6. When you add two negative integers, add the absolute values and keep the result negative, as shown in the following example:

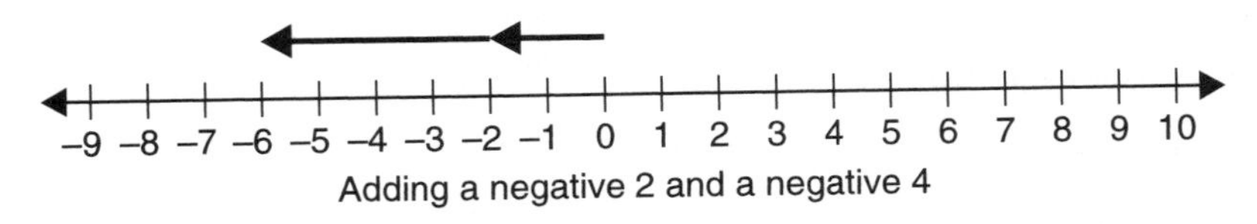

Adding a negative 2 and a negative 4

The number line is especially helpful to understand how to add two integers with different signs. For example, to add –7 + 5, start at zero, travel 7 units to the left because the 7 is negative, then travel 5 units to the right because the 5 is positive. As shown in the following example, the resultant arrow is at –2.

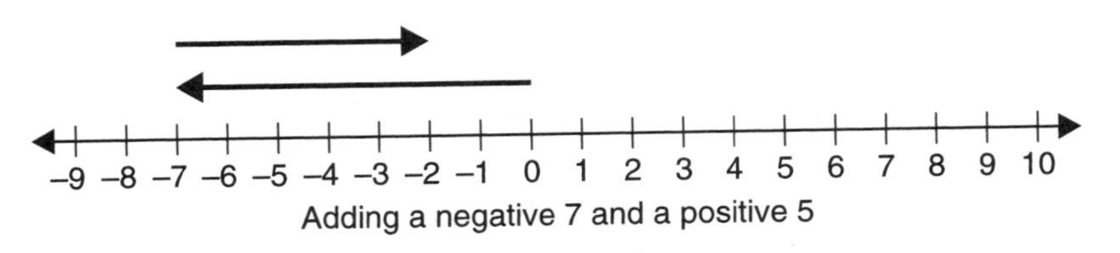

Adding a negative 7 and a positive 5

The sign of the sum will always be the sign of the integer with the highest absolute value. As seen on the number line, it is the integer with the longest arrow, or the biggest magnitude. The previous number line illustration shows that adding integers with different signs is simply finding the difference between the absolute values of the integers. To determine a difference, subtract. The important thing to understand when adding integers is that sometimes you add the absolute values (when both numbers have the same sign) and sometimes you subtract the absolute values (when the numbers have different signs).

Subtracting Integers

To subtract integers, have your child repeat the phrase "When you subtract, you add the opposite" several times until it becomes rote. When encountering subtraction of integers, first change subtraction to addition and then take the opposite of the second term (change it's sign). If you can help your child to understand that subtraction is the same as adding the opposite, then both addition and subtraction become similar processes. For example, $7 - 10$ is the same as $7 + -10$, and $-5 - -3$ is the same as $-5 + 3$.

The Rules for Addition and Subtraction of Integers

Here is the summary of the rules for adding and subtracting integers:

- If the problem is subtraction, first change the problem to add the opposite and then follow the guidelines for addition.
- If the integers have the same sign, add the absolute values of the integers and keep the sign.

- If the integers have different signs, subtract the absolute values of the integers and take the sign of the integer with the largest absolute value.

Multiplication and Division of Integers

Just as with fractional arithmetic, multiplication and division of integers is an easier process than addition and subtraction. To multiply or divide integers, multiply or divide the absolute values of the integers and follow these rules to determine the sign of the result:

- If the integers have the same sign, the resultant product or quotient is positive.
- If the integers have different signs, the resultant product or quotient is negative.

For example, $-2 \times 4 = -8$ because the two factors have different signs, $-10 \times -7 = 70$ because the two factors have the same signs, $\frac{20}{-4} = -5$ because the numbers have different signs, and $-60 \div -4 = 15$ because the numbers have the same sign.

Alternatively, you can interpret the rules for multiplication of several integers as follows:

- If there is an even amount of negative factors, the resultant product is positive.
- If there is an odd amount of negative factors, the resultant product is negative.

For example $(-2)^4 = -2 \times -2 \times -2 \times -2 = 16$ because there is an even amount of negative factors, while $(-2)^5 = -2 \times -2 \times -2 \times -2 \times -2 = -32$ because there is an odd amount of negative factors.

Looking Ahead

You child will encounter integer arithmetic in the future; for example, when she has to add polynomials, such as $-3x + 7 + 5x - 8 = (-3x + 5x) + (7 - 8) = 2x - 1$. Integer arithmatic will also come up when multiplying binomials, such as $(x - 3)(x + 5) = x^2 + 5x + -3x + -15 = x^2 + 2x - 15$, or working with negative fractions. Because the fraction bar means division, a negative fraction, such as $-\frac{2}{3}$, can be similarly expressed as $\frac{-2}{3}$ or $\frac{2}{-3}$.

Expressions and Equations

It is important for your child to distinguish between an expression and an equation. In early elementary school, your child was probably introduced to the word *solve*, which meant a variety of things, most importantly to find an answer in math. As your child progresses to pre-algebra and algebra, a distinction is made between an expression and an equation. In math, you evaluate and simplify expressions, and you solve equations. They are two different processes.

Expressions

An *expression* is a mathematical phrase that contains numbers, variables, and operations. Expressions consist of *terms*, which are the pieces that are separated by plus and minus signs. Some examples of expressions are $5n - 18$, which has two terms, or $2x^2 + 3x + 1$, which has three terms.

Evaluating Expressions

Expressions are often *evaluated,* which means to plug in, or substitute, a given number for the variable and then to simplify the result. To *simplify* an expression, on the other hand, means to put the expression in its simplest form. For example, the simplest form of the expression $5 + 3$ is 8. You can evaluate the expression $5n - 18$, when $n = 6$, by plugging in 6 for the variable n, meaning $5n - 18 = 5(6) - 18$. Then simplify to get $30 - 18 = 12$.

Looking Ahead

Expressions are often evaluated in higher mathematics. For example when finding the roots (solutions) of a quadratic equation of the form $ax^2 + bx + c$, you can use the expression on the right side of this formula: $x=\frac{-b\pm\sqrt{b^2-4ac}}{2a}$. You then substitute numerical values for a, b, and c to find the values of x that make the equation true. For example, to find the roots of the equation $x^2 + 5x + 6$, recognize that $a = 1$, $b = 5$, and $c = 6$. Make these substitutions to get $\frac{-5\pm\sqrt{5^2-4(1)(6)}}{2(1)}$. Evaluate the expression under the square root symbol using the order of operations: $\frac{-5\pm\sqrt{25-24}}{2(1)}$, then $\frac{-5\pm1}{2}$. This gives you the two solutions of $\frac{-5+1}{2}=\frac{-4}{2}=-2$ and $\frac{-5-1}{2}=\frac{-6}{2}=-3$.

Chapter 2 defines square roots in the vocabulary list, and formulas are explained later in this chapter.

Equations

An *equation* is a mathematical sentence that states that two expressions are equal. Equations contain an equal sign; expressions do not. Some examples of equations are $8n + 14 = 30$ or $10x - 5x - 12 = 23$.

Solving Equations

Equations are *solved*, which means to find the value of the variable. The goal when solving an equation is to *isolate the variable*, which means to get it alone, on one side of the equation. Solving an equation is a step-by-step process of making the equation simpler until the variable is alone on one side of the equation. This involves using the problem-solving strategy of working backwards. To work backwards, you use the concept of opposite operations. *Opposite operations* are operations that "undo" each other. Addition and subtraction are opposite operations; multiplication and division are opposite operations.

A simple example of solving an equation is to solve for $x - 14 = 3$. All that is required to solve this equation is to use the opposite operation of subtraction, which is addition, to get the variable x alone on the left side of the equation. In an equation, the two sides are said to be equal, so the concept of opposite operations is to "undo" operations on both sides of the equation so that the equality is maintained:

$x - 14 = 3$	**The given equation**
$x - 14 + 14 = 3 + 14$	**Undo subtraction by adding 14 to both sides.**
$x = 17$	**Combine like terms and isolate the variable.**

Equation solving can be a satisfying experience because you can check to see if your answer is correct. This can empower your child to feel confident when solving equations, and it is a good habit to practice right from the start with simple equations. Many teachers will require your child to check his answers, but even if they don't, you can give your child the motivation to perform this final step when solving equations.

To check a solution to an equation, plug in the answer for the variable in the original equation to see if it results in a true statement. If it does, then you can be confident that you were successful. If not, then you have to go back to recheck where an error may have occurred.

For example, to check the previous equation, plug in 17 for each occurrence of the variable x:

$x - 14 = 3$	**The original equation**
$17 - 14 = 3$	**Plug in your value of 17.**
$3 = 3$	**Your answer checks; your procedure was correct.**

The strategy of working backwards takes on additional steps when the equation is a bit more complicated and contains both addition and multiplication. For example, to work backwards with $3x + 5 = -10$, you must undo addition before you undo multiplication (the opposite method to order of operations).

$3x + 5 = -10$	**The given equation**
$3x + 5 - 5 = -10 - 5$	**Undo addition by subtracting 5 from both sides.**
$3x = -15$	**Combine like terms.**
$\frac{3x}{3} = \frac{-15}{3}$	**Undo multiplication by dividing both sides by 3.**
$x = -5$	**Isolate the variable.**

Again, perform a check to verify your work:

$3x + 5 = -10$	**The original equation**
$3(-5) + 5 = -10$	**Plug in your value of –5.**
$-15 + 5 = -10$	**Perform the integer multiplication.**
$-10 = -10$	**Perform the integer addition.**
	Your answer checks; your procedure was correct.

Looking Ahead

Solving equations is the cornerstone of most algebra work. In the future, your child will solve more complex equations, such as $8(x - 2) = 5x + 11$:

$8(x - 2) = 5x + 11$	**The given equation**
$8x - 16 = 5x + 11$	**Use the distributive property.**
$8x - 16 - 5x = 5x + 11 - 5x$	**Subtract 5*x* from both sides.**
$3x - 16 = 11$	**Combine like terms.**
$3x - 16 + 16 = 11 + 16$	**Add 16 to both sides.**
$3x = 27$	**Combine like terms.**
$\frac{3x}{3} = \frac{27}{3}$	**Divide both sides by 3.**
$x = 9$	**Isolate the variable.**

The distributive property was explained earlier in this chapter.

Graphing Points on a Coordinate Plane

A *coordinate plane* is the intersection of a horizontal number line, called the *x-axis*, and a vertical number line, called the *y-axis*. The two lines meet at a point that is called the *origin*. The lines make up four different areas on the grid, called *quadrants*. The quadrants are named quadrant I, II, III, and IV. As shown in the following figure, quadrant I is the upper-right section, and the quadrants are numbered in a counterclockwise direction.

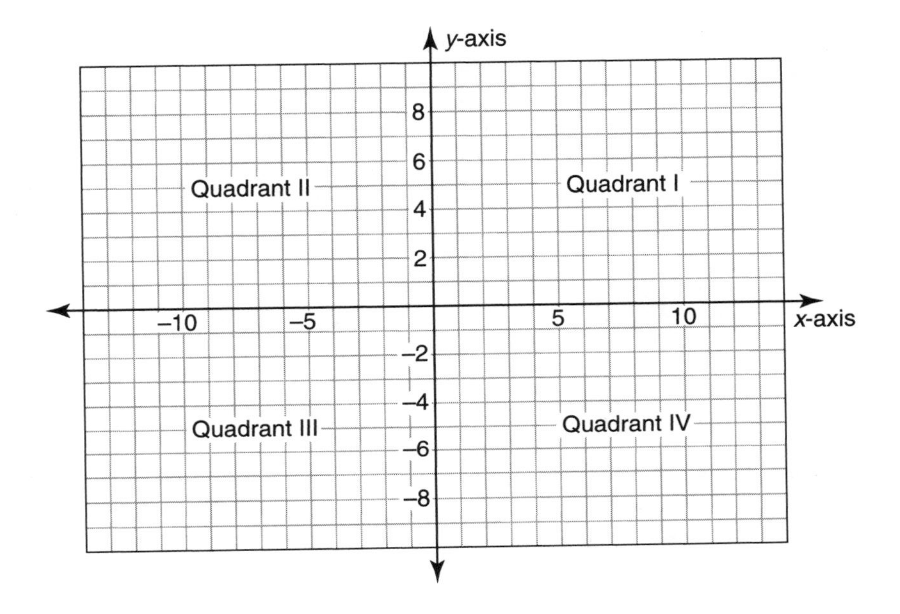

Points are identified on a coordinate plane by an ordered pair (x, y). The first value in the ordered pair is called the *x-coordinate*, and it is the distance that the point is from the y-axis, in either a right or left direction. If the point is to the right of the y-axis, the x-coordinate is positive. If the point is to the left of the y-axis, the x-coordinate is negative. The second value in the ordered pair is called the *y-coordinate*, and it is the distance that the point is from the x-axis, either above or below. If the point is above the x-axis, the y-coordinate is positive. If the point is below the x-axis, the y-coordinate is negative. The following figure shows the respective quadrants and the sign value of the x- and y-coordinates in each quadrant.

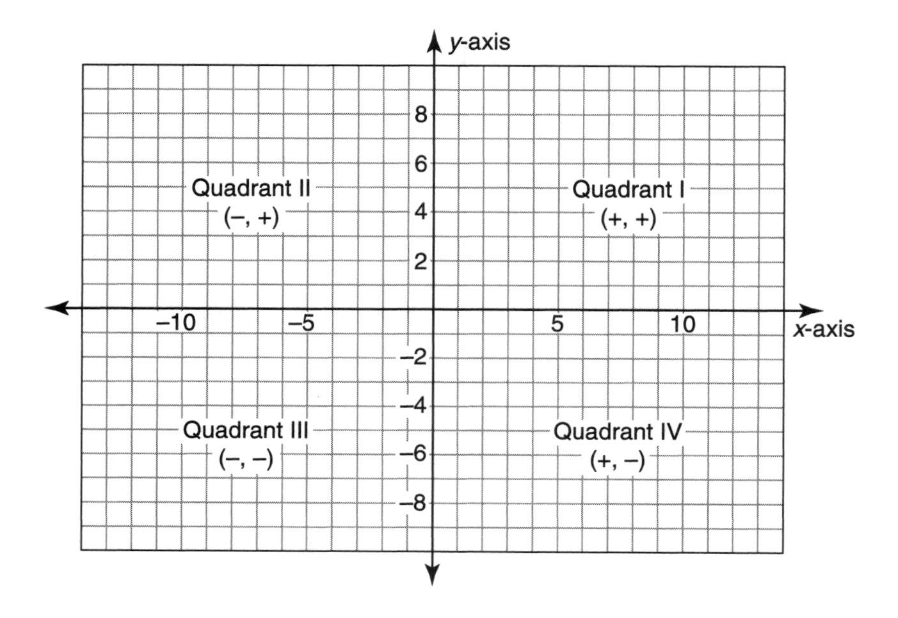

For example, study the points on the coordinate plane in the following figure. A description of the ordered pair for each point follows the figure.

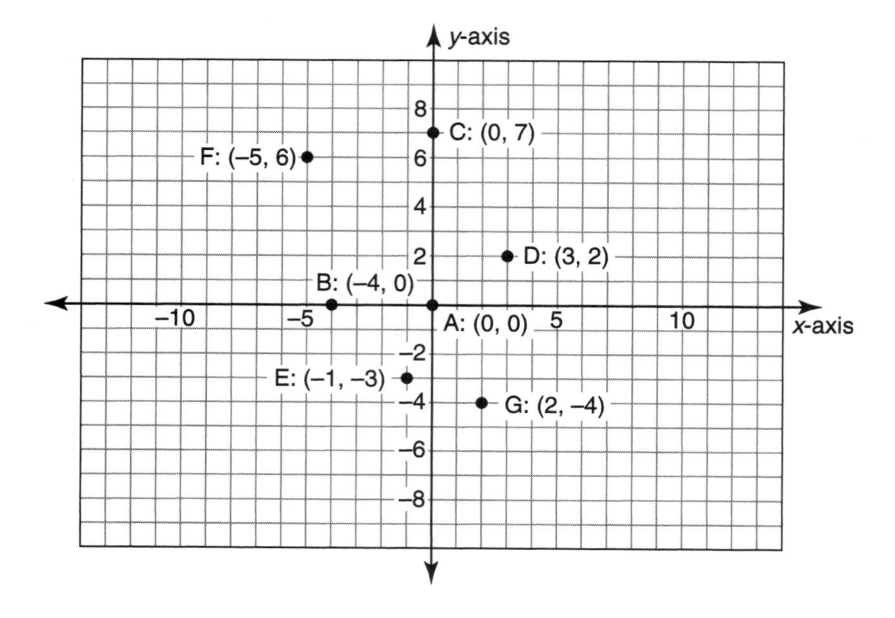

- Point A is at the origin, and its ordered pair is (0, 0).
- Point B is on the x-axis, 4 units to the left of the y-axis. Its ordered pair is (−4, 0).
- Point C is on the y-axis, 7 units above the x-axis. Its ordered pair is (0, 7).
- Point D is 3 units to the right of the y-axis and 2 units above the x-axis. Its ordered pair is (3, 2), and it is in quadrant I.
- Point E is 1 unit to the left of the y-axis and 3 units below the x-axis. Its ordered pair is (−1, −3), and it is in quadrant III.
- Point F is 5 units to the left of the y-axis and 6 units above the x-axis. Its ordered pair is (−5, 6), and it is in quadrant II.
- Point G is 2 units to the right of the y-axis and 4 units below the x-axis. Its ordered pair is (2, −4), and it is in quadrant IV.

Looking Ahead

Being able to plot points and graph functions on a coordinate plane is a very important skill in mathematics from middle school through college levels. A common element of algebra is graphing a line such as $y = 3x + 2$. This is an equation in slope-intercept, or $y = mx + b$, form. In this form, m (the value with x) is the slope of the line, and b is the y-intercept. The y-intercept is the place where the line crosses the y-axis. To graph this line, start at the y-intercept. Because b equals 2 in this equation, place a point on the point (0, 2) on the graph. From there, use the value of $m = 3$ as the slope. Write the slope in fraction form, so $3 = \frac{3}{1}$. Because slope is equal to $\frac{\text{rise}}{\text{run}} = \frac{\text{change in } y}{\text{change in } x}$, count up 3 units and over 1 unit to the right from (0, 2) and place a point at this location. Repeat this process to get at least five points and connect the points to form the line that represents the equation. The graph of the equation $y = 3x + 2$ is shown in the following figure.

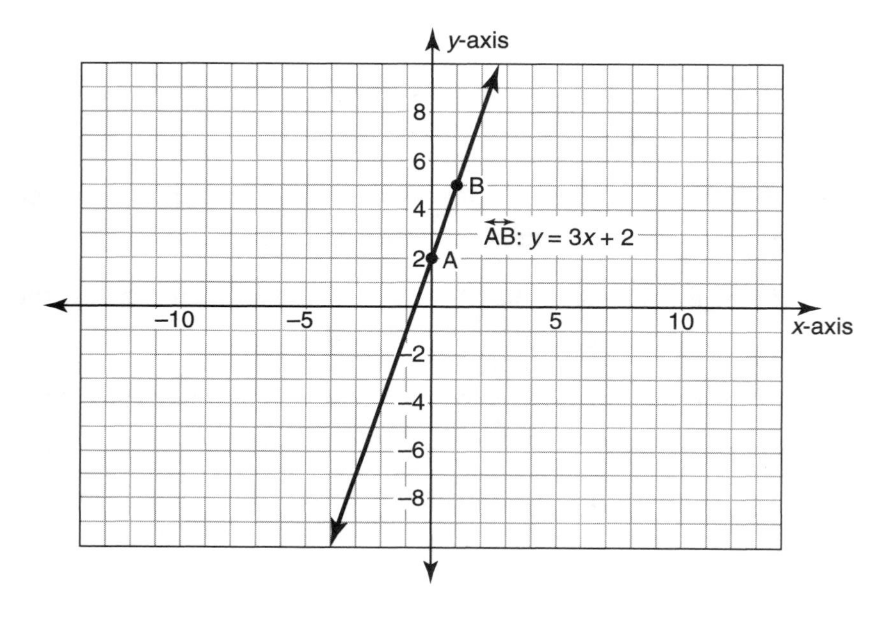

Using Formulas

Formulas are often encountered in math class from elementary school continuing into high school math and college study. Formulas also have an important role in science study, such as chemistry and physics. Financial applications use many formulas. A solid understanding of how to use formulas provides a good foundation for success in both academic study and in everyday life. A formula is simply an equation that has variables, usually more than one, and the variables have a specific meaning in relation to a geometric figure measure; a scientific measure; or a concept, such as money or interest rate in finance.

To use a formula, first determine which of the variables have known or given values and then either simplify the expression or solve the equation to find the unknown value. For example, you probably remember the familiar formula for the area of a rectangle; that is, area is equal to the product of the base times the height, written as $A = bh$. In the following figure, you are given the base length and the height length, and you can calculate the area.

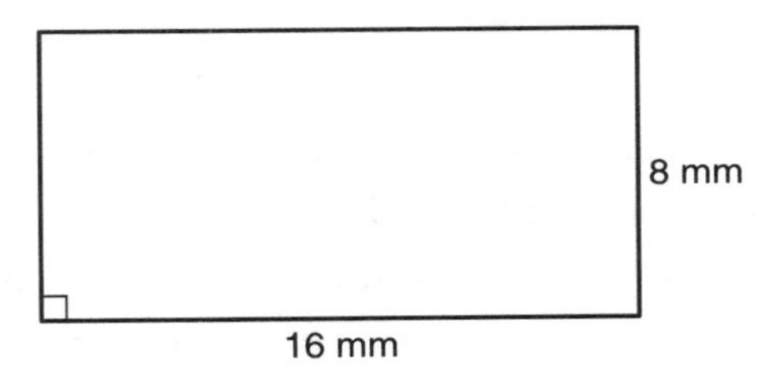

By referring to the figure, you determine that the base, b, is 16 millimeters and the height, h, is 8 millimeters. Plug these values into the area formula:

$A = bh$	**The given formula**
$A = 16 \times 8$	**Plug in the given values.**
$A = 128\ \text{mm}^2$	**Multiply and include the units.**

This example requires you to plug in values and then simplify. Sometimes, you substitute into a formula and must then solve an equation. Consider the familiar distance formula. $d = rt$, where d is the distance traveled, r is the rate of travel, and t is the time traveled. If you are given a problem in which you know that the car traveled 292.5 miles and the trip took 4.5 hours, you can use the formula and solve the equation to find the rate of travel:

$d = rt$	**The given formula**
$292.5 = r \times 4.5$	**Plug in the given values.**
$292.5 = 4.5r$	**Write the formula as an equation.**
$\frac{292.5}{4.5} = \frac{4.5r}{4.5}$	**Use opposite operations to divide both sides by 4.5.**
$65 = r$	**The solved equation**

The rate is 65 miles per hour.

Formulas are mathematical equations that involve measurement and, therefore, involve units of measure. Be especially careful to include the units in any answer. Also, before you use a formula, ensure that all the measurements are in consistent units before you use the formula. For example, the formula for the volume of a rectangular solid is $V = lwh$, where V is the volume, l is the length, w is the width, and h is the height. Look at the rectangular solid in the figure shown here and notice that the height is expressed in meters, while the length and width are in centimeters:

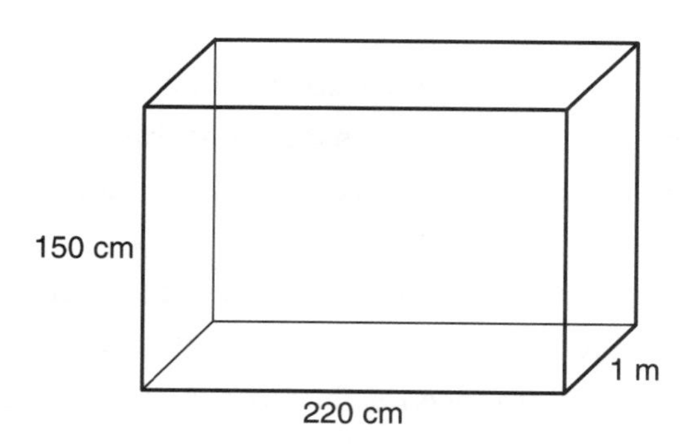

Before you use the formula, convert the measure of 1 m to 100 cm. Then proceed to find the volume:

$V = lwh$	**The given formula**
$V = 220 \times 100 \times 150$	**Substitute the given values.**
$V = 3{,}300{,}000 \text{ cm}^3$	**Multiply and include units.**

One of the most famous formulas is the Pythagorean theorem, which is used to find a missing measure of a right triangle. Given a right triangle with sides of a, b, and the side opposite to the right angle labeled c, the relationship among the sides is $a^2 + b^2 = c^2$.

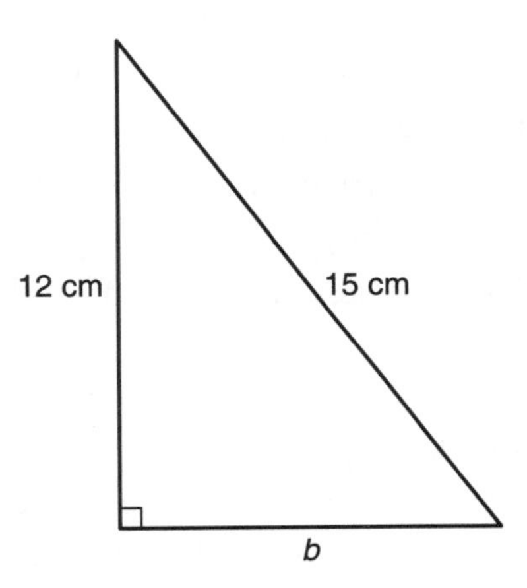

In the triangle in the figure, it is given that the side opposite to the right angle is 15 cm and one of the other sides is 12 cm. Use the formula and solve the resultant equation to find the missing side length:

$a^2 + b^2 = c^2$	**The given formula**
$12^2 + b^2 = 15^2$	**Substitute the given values.**
$144 + b^2 = 225$	**Evaluate the exponents.**
$144 + b^2 - 144 = 225 - 144$	**Subtract 144 from both sides.**
$b^2 = 81$	**Combine like terms.**
$\sqrt{b^2} = \sqrt{81}$	**Undo the exponent of 2, take the square root of both sides.**
$b = 9$ cm	**The measure of the missing side with units included**

Looking Ahead

Your child will be using formulas throughout high school math and science classes. In addition, everyday life requires formulas—from calculating the amount of paint needed to refurbish the walls in a room to determining the interest earned in a savings account. The formulas can get quite complicated in high school, but with the basic understanding of how to approach them, and how to evaluate expressions and solve equations, your child will find success in this area of mathematics.

For example, to find the distance between two points on a coordinate plane, you use the formula $d=\sqrt{(x_1-x_2)^2+(y_1-y_2)^2}$. To use this formula, you need to understand expressions, substitution, order of operations, and square roots, all the basics your child will learn prior to high school. The vocabulary list in chapter 2 defines square roots, and this chapter covers the other concepts. To find the distance between the points on the following coordinate graph, recognize that (x_1, y_1) can represent (3, 1) and (x_2, y_2) can represent (6, 5).

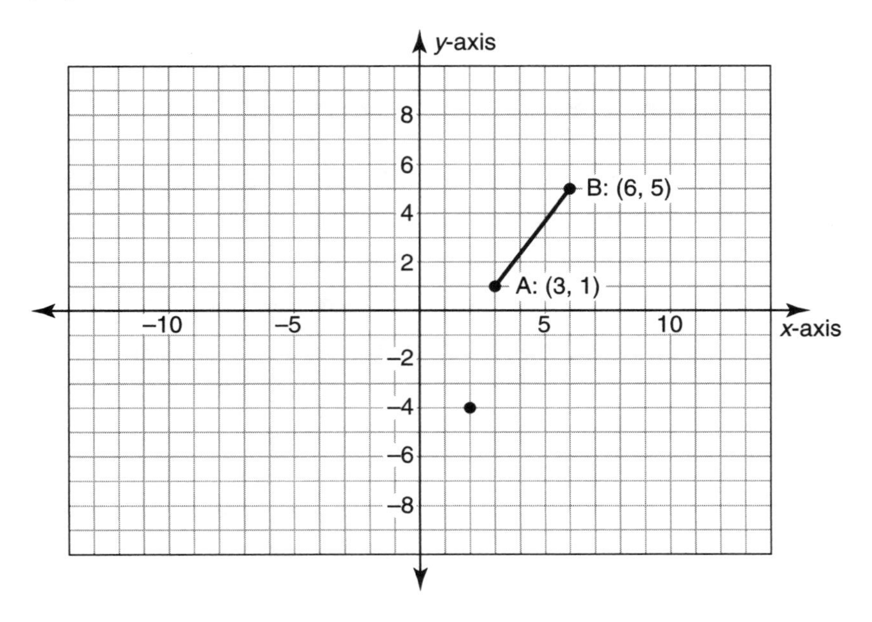

$d = \sqrt{(x_1 - x_2)^2 + (y_1 - y_2)^2}$	**The given formula**
$d = \sqrt{(3 - 6)^2 + (1 - 5)^2}$	**Substitute the given values.**
$d = \sqrt{(-3)^2 + (-4)^2}$	**Evaluate the grouping symbols.**
$d = \sqrt{9 + 16}$	**Evaluate the exponents.**
$d = \sqrt{25}$	**Add.**
$d = 5$ **units**	**Take the square root of the result.**

Chapter 4

Direction Decoding

This chapter concentrates on the often-confusing directions given with math problems. Sometimes the most difficult part of the whole problem is not solving the problem but figuring out exactly what the question is asking. By breaking directions down into smaller, easier to handle pieces of information, the problems become more clear and straightforward. This chapter provides specific examples of strategies that can be used when approaching math problems, such as looking for key terms and phrases, identifying the question, rephrasing the question, checking for reasonable answers, and working with diagrams. Use the techniques and suggestions with your child when solving math problems to ensure that she understands each question and correctly follows the directions given in each problem.

To implement these strategies, you should make it a point to get your child in the habit of reading the question a minimum of three times. Your child should read it the first time to get a sense of the question. Ask the student to rephrase the question in his own words to be sure he understands it. The second time the student should look for details, such as key phrases and numerical values. At this point, he should also check for any extra information. After solving the problem, the student should use the third read-though to check the answer to see if it makes sense and to be sure that the answer satisfies the question being asked. Of course, praise your child for his effort when working on complicated math problems. Remind the student that hard work pays off in the end!

Key Terms and Phrases

Math is a language that uses numbers and symbols, and common terms and phrases appear in math problems to represent the four basic operations and equivalence. For example, if the question is looking for "the sum of 16 and 23," the key word *sum* tells the student to add the two values. Because of this, relate to your child that translating these vocabulary words into symbols is just like translating from one language into another, like Spanish into English. Students should look for the key words and convert them into math symbols that can be worked with to find a final answer.

The following table summarizes many of these common terms in five categories.

+	−	×	÷	=
Add Sum Increased by More than Plus Exceeds	Subtract Difference Decreased by Less than Minus Reduced by	Multiply Product Of Times	Divide Quotient Into Split	Equal Result Is Total

This chapter, as well as later ones, will refer to this table numerous times. Review it with your child to help identify those important words that determine the operations and representations used to solve a problem.

Some specific examples of these key words in various instances are provided next. The step-by-step process should help you identify how common key terms are used in math problems and give specific examples of how to simplify the language by taking apart the words and phrases used in the problems. This way you, as the parent, can see how the strategies can be used to help your child better understand the vocabulary. Use the examples when helping your child break down complicated problems into smaller, more manageable pieces.

To change the given sentences into symbols, dissect each sentence into smaller pieces and replace these parts and vocabulary with the representation mathematical symbol. Underline or use a highlighter to point out key words and phrases (shown in italics here) that determine the operations used to solve a problem.

1. The product of 7 and 10 . . .
   product = multiply.
   7 and 10 = 7 × 10.
   The result of 7 × 10 = 70.
   The product of 7 × 10 is 70.

2. Ten less than 35 . . .
   less than = subtract.
   10 less than 35 = 35 – 10.
   The result of 35 – 10 = 25.
   Ten less than 35 is 25.

3. The quotient of 24 and 8 . . .
   quotient = divide.
   24 and 8 = 24 ÷ 8.
   The result of 24 ÷ 8 = 3.
   The quotient of 24 and 8 is 3.

4. Four more than a number . . .
   more than = add.
   4 more than = + 4.
   a number = a variable, n.
   Four more than a number is $n + 4$.

5. Three more than two times a number . . .
   more than = add.
   times = multiply.
   a number = a variable, n.
   two times a number = $2 \times n = 2n$.
   three more than = + 3.
   Three more than two times a number is $2n + 3$.

6. The sum of two numbers is 25. If the smaller number is 12, what is the value of the larger number?
   sum = add.
   smaller number = 12.
   unknown (larger number) = n.

$\mathbf{12 + n = 25}$	**Solve for *n*.**
$\mathbf{12 - 12 + n = 25 - 12}$	**Subtract 12 from both sides.**
$\mathbf{n = 13}$	**The larger number is 13.**

7. The difference of two numbers is 14. If the larger number is 20, what is the smaller number?
   difference = subtract.
   large number = 20.
   unknown (smaller) number = n.
   14 is the difference; therefore, $20 - n = 14$.

$20 - n = 14$	**Solve for *n*; isolate the variable.**
$20 - 20 - n = 14 - 20$	**Subtract 20 from each side.**
$-n = -6$	**Solve for a positive *n*.**
$-n \times -1 = -6 \times -1$	**Multiply each side by −1.**
$n = 6$	**The smaller number is 6.**

8. 5 times a number increased by 4 is equal to 29.
   5 *times* = multiply by 5.
   a number = a variable, n.
   increased by 4 = add 4.
   $5n + 4 = 29$.

$5n + 4 = 29$	**Isolate the variable by using opposite operations.**
$5n + 4 - 4 = 29 - 4$	**Subtract 4 from each side.**
$5n = 25$	**Simplify.**
$\frac{5n}{5} = \frac{25}{5}$	**Divide each side by 5.**
$n = 5$	**Isolate the variable.**

What Is the Question Asking, Anyway?

There are numerous strategies to help decipher directions and really get to the heart of the question being asked. First, help your child to read each question carefully and be sure that she is reading the entire question. An important consideration—especially if there are multiple sentences—is to remember that the actual question is usually contained in the last part of a problem and that it more than likely contains a question mark.

Working with your child to paraphrase a question into his own words will help you determine if your child really understands a problem. Have the student practice doing this each time he solves a problem. Discuss the meaning of the problem as your child sees it to help come up with a plan to solve the problem. Each of the problems shown in this section has an example of what each question may look like when your child rephrases it. Putting the question into terms more easily understood by your child will help him become more comfortable with creating a plan to answer it.

The examples show sample problems with explanations that include how to identify the particular question in each, how to rephrase each question, and ways to extract the important information to make a plan to solve the problem. Then, each question is examined again to check the solution.

1. Joe made 18 quarts of lemonade for the school picnic. How many gallons of lemonade did he make?

   4 quarts = 1 gallon

   First, identify the question. Look to the last sentence and notice the question mark.

   The question: How many gallons of lemonade did Joe make?
   Rephrased: Change quarts into gallons to find the total amount of lemonade.

   Identify the other information given in the problem:
   Joe made 18 quarts of lemonade for the picnic.
   Four quarts are equal to 1 gallon.

Now, use the information to solve the problem.

This question is looking for the number of gallons, and it gives you the number of quarts. To change from quarts to gallons, use the fact that every 4 quarts are equal to 1 gallon. Divide the total number of quarts, 18, by 4 to find the number of gallons: $18 \div 4 = 4.5$. There are 4.5, or $4\frac{1}{2}$, gallons in 18 quarts.

To complete the problem, have your child go back and read the question again to be sure that no other steps are necessary. This question is looking for the number of gallons, so the solution is 4.5 gallons.

2. A number increased by 3 is equal to twice the number. Write the equation that represents this statement.

First, identify the question. Look to the last sentence in the question and notice that there is no question mark in this problem. However, this sentence still contains the question being asked.

The question:	Write the equation that represents this statement.
Rephrased:	Change the words given in the problem into math symbols.

Identify the other information given in the problem:

A number is increased by 3.

This value is equal to twice the number.

Because the number is not given, this is represented by a variable, such as n.

Now, use the information to write the equation.

Translate the two parts of the first sentence of the problem into symbols.

A number	**represented by a variable, like *n***
***increased by* 3**	**add 3**
The first part of the sentence is:	**$n + 3$**
The second part of the sentence:	
is *equal to*	**=**
twice the number	**$2n$**

Putting the two parts together in order, the equation becomes $n + 3 = 2n$.

To complete the problem, have your child go back and read the question again to be sure that no other steps are necessary. This question is looking for the equation that represents the statement. Because this was translated to $n + 3 = 2n$, the question is answered; the equation does not need to be solved.

3. Shelly's cell phone plan is $25 per month plus $0.10 per minute used. If she only wants to spend a total of $30 or less each month on her cell phone, what is the greatest number of minutes she can use each month?

   First, identify the question. Look to the last sentence in the question and notice the question mark.

The question:	What is the greatest number of minutes she can use each month?
Rephrased:	What is the most minutes she can talk for $30 or less?

   Identify the other information given in the problem:
   The plan costs $25 per month.
   Each minute used costs an additional $0.10.
   She wants to spend a total of $30 or less each month.

   Now, use the information to answer the question.

This question is looking for the greatest number of minutes she can use, and you are given two different expenses. The plan costs $25 each month, so this is charged only one time each month. Because the total amount she wants to spend is $30, subtract $25 from $30 = $5. This is the amount left over to pay for the number of minutes she uses.

To find the number of minutes she can use, divide $5 by the cost of each minute.

$$\frac{5}{0.10} = 50$$

Five dollars divided by $0.10 is equal to 50, which means that 50 minutes costs $5.

To complete the problem, have your child go back and read the question again to be sure that no other steps are necessary. This question is looking for the greatest number of minutes she could use and not spend more than $30 on her monthly cell phone bill. Because 50 minutes costs $5 and the one-time monthly fee is $25, she would spend a total of $5 + $25 = $30 if she uses 50 minutes each month. Thus, the greatest number of minutes she can use each month is 50.

4. How many different ice cream cones can be made if there are two choices of size, three choices of ice cream, and two choices of toppings and only one selection is chosen from each category?

   First, identify the question. Look at the sentence in the question and notice the question mark. In this problem, the question is stated in the first part of the sentence.

The question:	How many different ice cream cones can be made?
Rephrased:	Find the total possible combinations of size, flavor, and toppings based on the choices given.

Identify the other information given in the problem:

There are 2 choices of size.
There are 3 choices of ice cream.
There are 2 choices of toppings.
Only 1 selection is chosen from each category.

Now, use the information to solve the problem.

This question is looking for the number of different ice cream cones. The total number of combinations can be found a number of ways. One of these ways is to make an organized list of the possibilities. To do this, use examples for each of the choices to be made. For example, use small (S) and large (L) for the two sizes. For the choices of ice cream, use flavors, such as vanilla (V), chocolate (C), and mint (M). Then, possible toppings could be a cherry (CH) or sprinkles (SP).

Make an organized list of all the different cones using these examples.

S-V-CH	small vanilla with a cherry
S-V-SP	small vanilla with sprinkles
L-V-CH	large vanilla with a cherry
L-V-SP	large vanilla with sprinkles
S-C-CH	small chocolate with a cherry
S-C-SP	small chocolate with sprinkles
L-C-CH	large chocolate with a cherry
L-C-SP	large chocolate with sprinkles
S-M-CH	small mint with a cherry
S-M-SP	small mint with sprinkles
L-M-CH	large mint with a cherry
L-M-SP	large mint with sprinkles

Because there are 12 different types of cones in the list, there are a total of 12 combinations. Making an organized list is a problem-solving strategy explained in even more detail in chapter 5. If your child works well with visual examples or likes to draw, consider having her use colored pencils or crayons to draw all the possible combinations!

Another way to find the total number of different possibilities is to multiply the total number of choices in each category. This is known as the *counting principle.* Because there are 2 choices of size, 3 choices of flavors, and 2 choices of toppings, multiply these choices: $2 \times 3 \times 2 = 12$. There are 12 different cones that can be made.

To complete the problem, have your child go back and read the question again to be sure that no other steps are necessary. This question is looking for the total number of different cones that can be made with the choices given in the problem. The total number of different cones is 12.

5. Pat leaves his house at 1:00 PM to ride his bike. If at 3:00 PM he has traveled 28 miles without stopping, what is his average rate of speed in miles per hour?

   First, identify the question. Look at the last sentence in the question and notice the question mark.

The question:	What is his average rate of speed in miles per hour?
Rephrased:	How fast is Pat going on his bike during the ride?

   Identify the other information given in the problem:

   Pat leaves at 1:00 PM and rides until 3:00 PM, so he rides for 2 hours.

   In 2 hours, he has traveled 28 miles.

   Now, use the information to solve the problem.

This question is looking for the rate of speed and gives you the time and the distance. Use the formula *distance = rate × time*, or $d = r \times t$. To solve the problem, substitute the given values into the formula.

$d = r \times t$	**The formula**
$28 = r \times 2$	**Substitute into the formula.**
$28 = 2r$	**Simplify.**
$\frac{28}{2} = \frac{2r}{2}$	**Divide each side by 2 to get the variable alone.**
$14 = r$	**Isolate the variable.**

To complete the problem, have your child go back and read the question again to be sure that no other steps are necessary. This question is looking for the rate of speed, which is equal to r in the formula $d = r \times t$. Because $r = 14$, then Pat's rate of speed was 14 miles per hour.

Did I Answer the Question Being Asked?

Another area that causes student errors is when students can solve the problem but do not answer the question at hand. When an answer is reached, be sure that your child is actually answering the question. For example, a geometry question could ask a student to find the value of x in a figure or ask for the measure of each angle in the diagram. The latter would require the student to find the value of x but also requires an extra step or two to find the measure of each angle to complete the problem.

The following problems illustrate questions where students need to be careful about the final answer they give. Work with your child on the strategies and steps to follow to make sure he is answering the question.

1. Celine has two binders of trading cards. The second binder has 30 more cards than the first binder. She has a total of 80 cards in both binders. Based on this information, the following equation can be used to find the number of cards, x, in the first binder.

$$x + x + 30 = 80$$

How many cards are contained in the second binder?

First, identify the question. Look to the last sentence in the question below the equation and notice the question mark.

The question:	How many cards are contained in the second binder?
Rephrased:	What is the total number of cards in the binder with more cards in it, when the total of both binders is 80 cards?

Identify the other information given in the problem. Use the information in the question itself as well as the given equation. In a question such as this, is it critical that all of the information be noted. This information gives the background to the problem and explains how the equation was set up.

Celine has two binders.
One binder has 30 more cards than the other.
The total number of cards in both binders is 80.
The x in the given equation represents the number of cards in the first binder.
The given equation to find x is $x + x + 30 = 80$.

Now, use the information to solve the problem.

Use the given equation to solve for x. Then use this value to find the number of cards in the second binder.

$x + x + 30 = 80$	**Write the equation.**
$2x + 30 = 80$	**Combine like terms.**
$2x + 30 - 30 = 80 - 30$	**Subtract 30 from each side.**
$2x = 50$	**Simplify.**
$\frac{2x}{2} = \frac{50}{2}$	**Divide each side by 2 to get the variable alone.**
$x = 25$	**Isolate the variable.**

To complete the problem, have your child go back and read the question again to be sure that no other steps are necessary. This question is looking for the number of cards in the second binder. The value of x represents the number of cards in the first binder, so one must add 30 to this value to find the amount in the second binder: $25 + 30 = 55$. There are 55 cards in the second binder.

For a quick check, have your child use the value of x in the equation to see if the equation is true. The equation is $x + x + 30 = 80$, so substitute to get $25 + 25 + 30 = 80$, which is true. Another way to check to see if an answer is reasonable is to take the values for each binder see if they total 80. Because the first binder has 25 cards and the second binder has 55 cards, the total is $25 + 55 = 80$, which also checks.

A student could be tempted in this type of question to go directly to the given equation and ignore the other information in the question. Just solving for x in this problem does not satisfy the question; the student needs to find the number of cards in the binder with 30 more cards in it. For this to be clear and to be sure that the question at hand is answered, the reader needs to understand clearly the information given.

2. The cost of a skating lesson is $12.50, and skate rental is $3. What is the total cost of 6 lessons with skate rental?

First, identify the question. Look at the last sentence in the question and notice the question mark.

The question:	What is the total cost of 6 lessons with skate rental?
Rephrased:	How much money is needed for 6 lessons and 6 skate rentals?

The final answer in this question must be the total amount of money needed for 6 lessons, each of which includes skate rental.

Identify the other information given in the problem:

The cost of one skating lesson is \$12.50.
The cost of renting skates during a lesson is \$3.
There is a total of 6 lessons.

Now, use the information to solve the problem.

There are a number of approaches to a problem such as this one. One approach is to find the total amount for one lesson and one rental and then multiply this value by the number of lessons to be taken. Because each lesson is \$12.50 and each rental is \$3, the total cost for one lesson and one rental is \$12.50 + \$3 = \$15.50. Now, multiply this amount by 6 to find the total cost of the 6 lessons with skate rental: \$15.50 × 6 = \$93.

Another method is to find the total cost of 6 lessons and add it to the total cost of 6 skate rentals. The total cost of 6 lessons is \$12.50 × 6 = \$75, and the total cost of 6 skate rentals is \$3 × 6 = \$18. Thus, the final cost with rentals is \$75 + \$18 = \$93.

To complete the problem, have your child go back and read the question again to be sure that the question is answered. In a question such as this, a child could make the common mistake of finding the total for one lesson with skate rental but forget to find the total for six lessons with skate rental. If your child gave a final answer of \$15.50, this is most likely the case. Be sure that your child answers the question being asked. Reading the question a final time could be the difference between a correct and incorrect solution.

3. Two angles are supplementary. The measure of one angle is twice the measure of the other angle. What is the measure of the larger angle?

First, identify the question. Look at the last sentence in the question and notice the question mark.

The question:	What is the measure of the larger angle?
Rephrased:	When finding two angles, how many degrees are in the bigger angle?

Identify the other information given in the problem:

There are two angles in the problem.

The angles are supplementary, so the sum of their measures is 180 degrees.

The larger angle is twice the measure of the smaller one, or double in measure.

Now, use the information to solve the problem.

To solve this question, take the values given in the figure and write an equation based on the given information. Because one angle is twice the measure of the other angle, use x to represent the smaller angle and $2x$ to represent the larger angle. Then, because the angles are indicated by x and $2x$ and they are supplementary, write an equation that adds x and $2x$ and sets it equal to 180 degrees.

$x + 2x = 180$	**Write the equation.**
$3x = 180$	**Combine like terms.**
$\frac{3x}{3} = \frac{180}{3}$	**Divide each side by 3 to get the variable alone.**
$x = 60$	**Isolate the variable.**

To complete the problem, have your child go back and read the question again to be sure that no other steps are necessary. A common mistake in this type of question is to solve for the value of x, when the question asks for the measure of the larger angle. Because the value of $x = 60$, this value must be substituted into the expression to find the measure of the larger angle. Therefore, $2x = 2(60) = 120$ degrees. Be sure that your child answers the question being asked, and doesn't just solve for the value of x if more steps are necessary.

A great way to check to see if the question has been answered is to draw a picture, where appropriate, to see if the answer "fits" in the question. In this problem, the two angles are supplementary, which means that the sum of the degrees of the angles is 180 degrees. Therefore, draw a picture of two supplementary angles and fill in the values found when solving the problem. A possible diagram appears here.

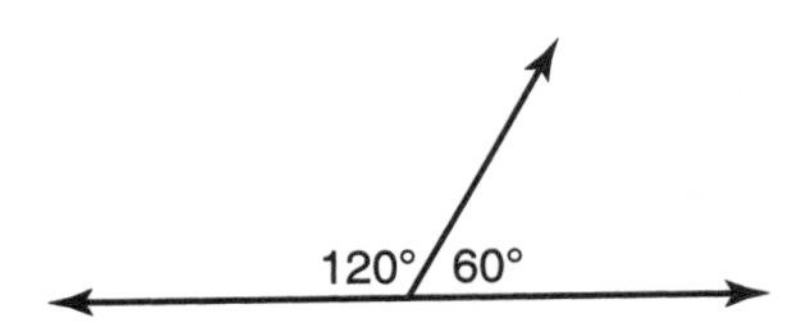

The diagram matches the degree measures found in the question, so the final answer is the larger angle, 120 degrees. Using a picture or diagram to help solve a problem is explained in further detail in chapter 5.

In addition, a question of this type could also be solved by "guess and check." To use this method, try two values, where one is double the other, and check to see if they add up to 180 degrees. For example, try 50 and 100. The sum of these numbers is 150, which is too small. Try 70 and 140. These numbers add to 210, which is too large. Now, try something in between, such as 60 and 120, which would give a sum of 180 degrees—the number of degrees in a straight line. Thus, the larger angle is 120 degrees.

4. Ten less than five times a number is 40. What is the value of triple the number?

   First, identify the question. Look at the last sentence in the question and notice the question mark.

The question:	What is the value of triple the number?
Rephrased:	Find the number first, and then multiply this value by three to find triple the number.

   Identify the other information given in the problem:

   The first part of the sentence is "Ten less than five times a number."
   This value is equal to 40.
   Because the number is not given, this is represented by a variable, such as n.
   The final answer will be the value of triple the number, not just the number itself.

   Now, use the information to write the expression.

   Translate the two parts of the first sentence of the problem into symbols.

This question involves a *number*.	**a variable, n**
Ten *less than*	**subtract 10 from a value, or – 10.**
Five *times* a number	**$5n$**
The first part of the sentence is:	**$5n - 10$**
The second part of the sentence, is *equal to* 40	**= 40**

   Putting the two parts together in order, the equation becomes $5n - 10 = 40$.

Solve this equation to find the value of the number.

$5n - 10 = 40$	**Write the equation.**
$5n - 10 + 10 = 40 + 10$	**Add 10 to both sides.**
$5n = 50$	**Simplify.**
$\frac{5n}{5} = \frac{50}{5}$	**Divide each side by 5 to get the variable alone.**
$n = 10$	**Isolate the variable.**

Because $n = 10$, the number is equal to 10.

To complete the problem, have your child go back and read the question again to be sure that no other steps are necessary. This question is looking for the value of triple the number, which is the number multiplied by 3. The number is, therefore, $10 \times 3 = 30$. The final answer to the question is 30. Be sure that your child reads the question multiple times to ensure she answers the question being asked, as in this question.

Does My Answer Make Sense?

Another area of difficulty for students is making sure that their answers are reasonable. In many cases, students can figure out the correct operations to solve a problem, but the order in which the calculations should be done is not obvious. This can lead to solutions where the answers do not make sense. If these answers are checked at the completion of the problem and unreasonable solutions are found, a student can then go back into the process and try a different approach that may yield a more sensible answer.

Always have your child check the final answer to make sure it is reasonable. If the value reached for the solution does not make sense, then go back and check the key words and phrases. Try a different approach to the question or a different order in the calculations using the values stated in the problem.

1. A school bus can carry 64 students. At Maple Hills Elementary School, there are 8 fifth-grade classes, and each class has 22 students. If the entire fifth grade is going on a field trip, how many school buses are needed?

   First, identify the question. Look at the last sentence in the question and notice the question mark.

The question:	How many buses are needed?
Rephrased:	What is the number of buses needed for the entire fifth grade?

   Identify the other information given in the problem:
   A school bus can carry 64 students or fewer.
   The students are from Maple Hills Elementary School.
   There are 8 fifth-grade classes.
   Each class has 22 students.
   All of the fifth-graders are going on the field trip.

   Now, use the information to answer the question.

   This question is looking for the number of buses needed to take the entire fifth grade on a field trip. To do this, first find the total number of students. Because there are 22 students in a class and a total of 8 classes, multiply these values to find the total number of fifth graders.

   $$8 \times 22 = 176 \text{ students}$$

   Each bus can hold a total of 64 students. Divide the total number of students by 64 students on each bus.

   $$176 \div 64 = 2.75$$

According to this information, 2.75 buses are needed to transport all of these students.

However, to complete the problem, have your child go back and read the question again to be sure that no other steps are necessary. This question was looking for the number of buses that would be needed to take all of the fifth-graders on a field trip. The answer of 2.75 buses is not a reasonable answer because you cannot drive three-quarters of a bus—the number of buses needs to be a whole number. This amount needs to be rounded to 3 buses to transport all of the students; 2 buses would not be enough. Multiply 3 buses by the total number of students that can ride on each bus, 64, to check this solution: $3 \times 64 = 192$, which leaves plenty of room for all of the students.

2. Brad is six years older than Andrea. The sum of their ages is 70. How old is Andrea?

   First, identify the question. Look at the last sentence in the question and notice the question mark.

   The question: How old is Andrea?
   Rephrased: Comparing Brad to Andrea, find Andrea's age.

   Identify the other information given in the problem.
   Brad is six years older than Andrea.
   The sum of their ages is 70.
   Sum is a key word for addition.

   Now, use the information to answer the question.

   This question is looking for Andrea's age. Use the comparison to Brad's age to help find this value. One way to solve this question is to use the strategy of guess and check. For example, because Brad is six years older, take two numbers that are six apart, such as 30 and 36. Now, add these values to see if the sum is 70.

$$30 + 36 = 66$$

This is not the correct sum and is slightly smaller than 70, so try two values larger than these on the next guess, such as 34 and 40. Be sure, however, that the two numbers that represent their ages are six apart.

$$34 + 40 = 74$$

This sum is too high, so on the third guess, try values between the ones already used, such as 32 and 38.

$$32 + 38 = 70$$

These numbers, 32 and 38, are six numbers apart and also have a sum of 70.

However, to complete the problem, have your child go back and read the question again to be sure that no other steps are necessary. This question was looking for Andrea's age. Because Brad is six years older than Andrea, 38 represents Brad's age, and 32 represents Andrea's age. The correct answer to this question is 32, not 32 + 38. Be sure that your child checks the solution and that he does not give Brad's age or both ages as a final answer.

3. The three angles of a triangle are represented by x, $x + 10$, and $x + 20$. What is the measure of the largest angle of the triangle?

First, identify the question. Look at the last part of the sentence in the question and notice the question mark.

The question: What is the measure of the largest angle of the triangle?

Rephrased: Find the value of x and then use this value to find the largest angle as a final answer.

Identify the other information needed to solve the problem:

There are three angles in any triangle.
The sum of their measures is always 180 degrees.
One angle is labeled x, another is $x + 10$, and the third is $x + 20$.

Now, use the information to solve the problem.

To solve this question, take the expressions given for each angle. Then, write an equation that adds these three angles and sets the sum equal to 180 degrees.

$x + x + 10 + x + 20 = 180$	**Write the equation.**
$3x + 30 = 180$	**Combine like terms.**
$3x + 30 - 30 = 180 - 30$	**Subtract 30 from both sides.**
$3x = 150$	**Simplify.**
$\frac{3x}{3} = \frac{150}{3}$	**Divide each side by 3.**
$x = 50$	**Isolate the variable.**

To complete the problem, have your child go back and read the question again to be sure that no other steps are necessary. A common mistake in this type of question is to solve for the value of x and stop, when the question asks for the measure of the largest angle. The value of $x = 50$, so now calculate the other two angles. The angle represented by $x + 10 = 50 + 10 = 60$. The angle represented by $x + 20 = 50 + 20 = 70$. Thus, the three angles are 50, 60, and 70 degrees, the largest of which is 70, so 70 degrees is the final answer.

A good way to check a problem that involves the angles of a triangle is to add the values to see if all three angles add to 180 degrees. Because $50 + 60 + 70 = 180$, and 60 is 10 more than 50 and 70 is 20 more than 50, this problem checks. If your child got values that did not add to 180 degrees, the solution would be incorrect.

As in this question and any other, be sure that your child answers the question being asked, not just solves for the value of x if more steps are necessary.

4. When flipping a coin two times, what is the probability of getting tails both times?

First, identify the question. Look at the last part of the sentence in the question and notice the question mark.

The question:	What is the probability of getting tails both times?
Rephrased:	When flipping the same coin twice, how likely is it that tails will come up two times in a row?

Identify the other information given or known about the problem:

There is only one coin.
The coin has two sides, heads and tails.
The coin is flipped two times, one right after the other.

Now, use the information to answer the question.

This question is looking for the probability that the same coin will come up tails when flipped two times. The probability of an event is expressed as follows:

$$P(E) = \frac{\text{number of ways an event can occur}}{\text{total number of possibilities}}$$

Because there are two sides to each coin and tails is on one of the sides, the probability is equal to $P(\text{tails}) = \frac{1}{2}$.

This question asks for the probability of getting tails two times. Therefore, if the probability of getting tails once is $\frac{1}{2}$, then multiply the probability of the events together. The probability is equal to $P(\text{tails twice}) = \frac{1}{2} \times \frac{1}{2} = \frac{1}{4}$.

However, many times in questions such as these, it may be difficult for your child to determine whether to multiply the values or add them together. Use a check for reasonable answers to help with this situation. If the probabilities were incorrectly added together in this case, the probability of getting tails twice would equal $\frac{1}{2} + \frac{1}{2} = \frac{2}{2} = 1$. A probability of 1, or one whole, would indicate that this event would always occur; this is known as a *certain event*. However, there are clearly other possibilities that could happen, such as getting heads twice, getting heads first and tails second, or getting tails first and heads second. This would make a probability of 1 for this problem unreasonable.

To complete the problem, have your child go back and read the question again to be sure that no other steps are necessary. This question was looking for the probability of getting heads both times when flipping a coin twice. The reasonable solution to this problem is $\frac{1}{4}$.

What About Questions That Have Diagrams?

Certain problems that include a diagram contain information both above and below the figure. These questions are commonly known as *split-graphic questions.* Be sure that if a picture is included in the problem, that your child carefully reads the information before the graphic as well as below the graphic so that no information is overlooked. Don't forget the information given above the diagram or picture; it is most likely necessary to solve the problem. On the other hand, if a diagram is not given, especially in a geometry question, draw one. Seeing the relationships in a picture is usually very helpful in setting out to solve a problem.

An additional problem area is extra information. Many times, questions contain information that is not necessary to solving the problem, but students feel they should use all the information presented in the question. Double-check that your child only uses information necessary to solve the problem.

Use the following questions to help model the strategies and tips for solving problems that contain figures and diagrams within the question.

1. The length of a rectangular yard is 20 meters, and the width is 10 meters. What is the perimeter of the yard?

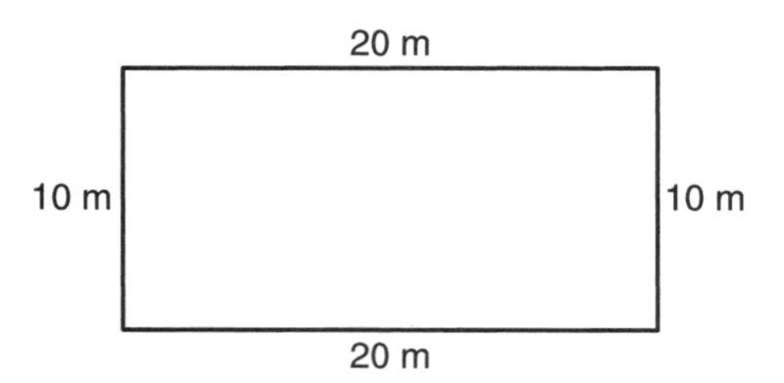

First, identify the question. Look at the last sentence in the question and notice the question mark.

The question:	What is the perimeter of the yard?
Rephrased:	What is total distance around the yard?

Identify the other information given in the problem, especially the picture:

The yard is in the shape of a rectangle.
The length of the yard is 20 meters.
The width of the yard is 10 meters.

Now, use the information to solve the problem.

This question is looking for the perimeter of the yard. The perimeter of a figure is the distance around the figure, or the sum of the sides of the figure. Use the formula *perimeter* = *length* + *width* + *length* + *width*, or $p = l + w + l + w$. To solve the problem, substitute the given values into the formula.

$p = l + w + l + w$	**Perimeter formula**
$p = 20 + 10 + 20 + 10$	**Substitute into the formula.**
$p = 60$	**Add.**

To complete the problem, have your child go back and read the question again to be sure that no other steps are necessary. This question is looking for the perimeter of the yard, which is equal to p in the formula $p = l + w + l + w$. Because $p = 60$, the perimeter of the yard is 60 meters. A common mistake here is for the student to use only two of the sides to find the perimeter. For example, if an answer of 30 meters was given, more than likely the student only added the two given values of 10 and 20, when actually two lengths and two widths must be added.

Because perimeter is the distance around an object, imagine taking a walk around a yard such as this one. You would walk 20 meters, then 10 meters, then 20 meters, then 10 meters if you walked the perimeter. This distance is 20 + 10 + 20 + 10, or 60 meters.

Another common error is for the student to multiply the given dimensions in the problem, which would incorrectly find the area of the rectangle. If an answer of 200 was given, the student probably multiplied 10×20 to find the area, not the perimeter, of the rectangle.

2. The following Venn diagram compares the number of students on the track team and the number of students on the swim team.

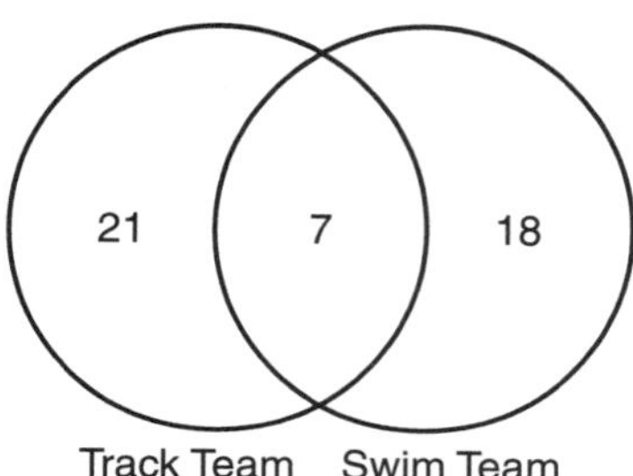

Based on the figure, how many students are on the track team?

First, identify the question. Look at the last sentence in the question under the graphic and notice the question mark.

The question:	How many students are on the track team?
Rephrased:	What is the total number of students on the track team, including those that are on both teams?

Identify the other information given in the problem. Use the information preceding the diagram and the information in the diagram, as well as in the question itself.

The figure compares the number of track team members with the number of swim team members.

There is a 21 in the part of the track team circle that does not overlap with the swim team circle.

There is an 18 in the part of the swim team circle that does not overlap with the track team circle.

There is a 7 in the part of the figure where the two circles overlap.

Now, use the information to solve the problem.

The circles compare the number of students on two different teams. The number in the part of the circle for the track team that does not overlap with the circle for the swim team represents the number of students who are on the track team only. These students are not also on the swim team. In the same way, the number in the part of the circle for the swim team that does not overlap with the circle for the track team represents the number of students who are on the swim team only. These students are not also on the track team. The number that is in the overlapping section represents the number of students who participate on both teams.

Therefore, 21 students are on the track team only, 18 students are on the swim team only, and 7 students are on both teams. To find the total number of students on the track team, add the 21 students on the track team only with the 7 students on both teams. The total number of students on the track team is 21 + 7, or 28.

To complete the problem, have your child go back and read the question again to be sure that it has been answered. A student could be tempted in this type of question to use one of the given values in the figure as the solution, such as 21, so check to be sure that your child answers the question being asked. This question was looking for the total number of students on the track team, so the correct answer is 28, which was not a value given in the figure. If a student answered 21, she would have found the number of students on the track team that do not participate on the swim team, not the total number on the team.

3. In triangle ABC, the measure of angle B is 40 degrees and $m\angle A = m\angle C$.

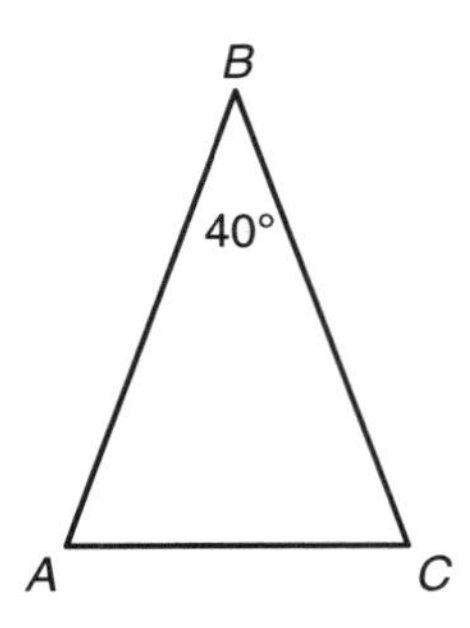

What is the measure of angle A?

First, identify the question. Look at the last sentence in the question under the graphic and notice the question mark.

The question:	What is the measure of angle A?
Rephrased:	Using the information given, find the number of degrees in angle A.

Identify the other information given in the problem. Use the information in the figure as well as that the question itself. Read the information preceding the figure, also, and mark it on the triangle in the figure.

There are three angles.

The sum of the measures of the interior angles of any triangle is 180 degrees.

The measure of angle B is 40 degrees.

The measure of angle A is equal to the measure of angle C.

Now, use the information to solve the problem.

To solve this question, take the information given about the three angles. Because the measure of angle B is 40, subtract $180 - 40 = 140$. This is the number of degrees in both angles A and C together. It is known that these two angles are equal; therefore, divide 140 by 2 to get the number of degrees in each angle: $140 \div 2 = 70$.

To complete the problem, have your child go back and read the question again to be sure that no other steps are necessary. A common mistake in this type of question is to ignore the information given in the sentence above the figure. Many times, students will go directly to the diagram to solve a problem while missing this necessary information. If it was not realized that angles A and C had the same measure, this problem would have been impossible to solve.

The student should now use the three angles found in the triangle to check his solution. The sum of the angles is $40 + 70 + 70 = 180$, and the measure of angle A is the same as the measure of angle C. Therefore, the answer checks.

4. A garden in the shape of a trapezoid is shown in the figure.

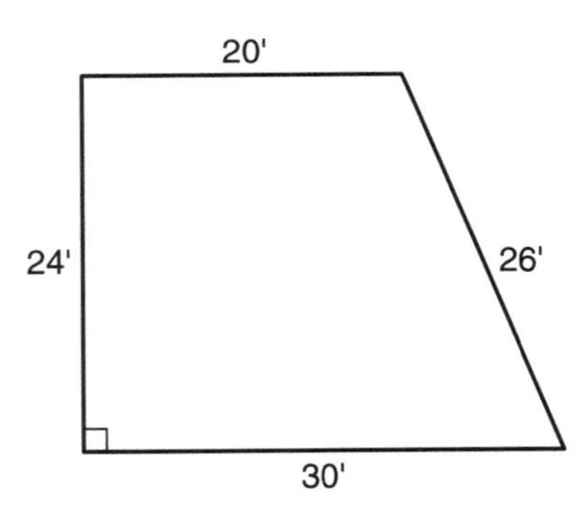

What is the area of this garden in square feet?

First, identify the question. Look at the last sentence in the question below the diagram and notice the question mark.

The question:	What is the area of the garden in square feet?
Rephrased:	Using the shape of the garden, find the number of square units that would cover this area.

Identify the other information given in the problem. Use the information preceding the figure and the information in the diagram, as well as that in the question itself.

The garden is in the shape of a trapezoid.
A trapezoid has four sides.
Two sides are parallel and uneven; the other two sides are not parallel and also uneven.
Area is the amount of square units that covers a region.
The units in the figure are feet.
The bases are labeled 20 and 30 feet, the sides are 24 and 26 feet.

Now, use the information to solve the problem.

The formula for the area of a trapezoid is based on the fact that *Area* = *base* × *height*. However in trapezoids, the bases are two different sizes. Therefore, find the average size base by adding the two bases and dividing this sum by 2. Then, multiply this average by the height of the figure. This measurement can be expressed as the formula below:

$A = \frac{1}{2}(b_1 + b_2)h$, where A is the area, b_1 and b_2 are the two bases, and h is the height of the trapezoid.

Substitute the values given in the figure into this formula to calculate the area. Be sure to watch for any extra information labeled in the figure, as all the values are not needed in this case.

$A = \frac{1}{2}(b_1 + b_2)h$	**Write the formula.**
$A = \frac{1}{2}(20 + 30) \times 24$	**Substitute into the formula.**
$A = \frac{1}{2}(50) \times 24$	**Add inside the parentheses.**

$A = \frac{1}{2}(1,200)$	**Multiply.**
$A = 600$	**Take $\frac{1}{2}$ of 1,200 to complete the process.**

To complete the problem, have your child go back and read the question again to be sure that no other steps are necessary. A student could be tempted in this type of question to use an extra value in the problem and answer the question incorrectly. The extra information given was the side labeled 26 feet. Because this side is not perpendicular to the bases, it cannot be used as the height of the trapezoid and is not needed to find the area. Also, this question is looking for the area of the figure, so be sure that your child gives a solution of 600 square feet, not just 600 feet.

What Have We Learned About Direction Decoding?

This chapter provides numerous strategies and techniques for parents to use with their child to break down the complicated directions of many math problems. Keep in mind that by simplifying directions and filtering out the important information, difficult math problems can be conquered. Remember to have your child restate the question in her own words to check if she really understands what the question is asking and what she needs to solve for. This needs to be done before a student can make a plan to solve the problem.

Help your child to extract the information from each question by looking for key words and phrases that usually define which operations are used in a problem. Also, have the student make a list of the information given in a question, including the information from pictures and diagrams. This technique will assist him in locating any extra information that is not necessary when solving the problem.

Remind your child to read the problem at least three times, always checking a final time to be sure that question is correctly answered. In addition have her check each solution to make sure that the answer makes sense. If the student is working through a problem that asks for the average height of students in a class and her answer is 10 feet, we know there has probably been a miscalculation somewhere!

Approach each question with a positive attitude when working with your child. If practiced, the strategies and techniques explained in this chapter will become a part of how your child solves math problems and help build confidence and success in math.

Chapter 5

Little Pieces Lead to Big Problems

Chapter 3 covered the rules you should know, and chapter 4 tackled direction decoding. This chapter takes these ideas and expands upon them to give you the skills and confidence to help your child conquer math difficulties. There are no really big problems in mathematics, just problems that consist of many little pieces. By keeping this perspective, you will do a great deal to sooth math anxiety both for yourself and your child.

Keep in mind that in learning math strategies, you are helping your child to develop good problem-solving skills for many aspects of life. Two key ingredients are hope and success. Hope is the emotional feeling you need to get started. After working through chapters 2, 3, and 4, you and your child should be confident that a solution can always be reached. Any past successes are reminders that help your child to persevere through the current problem. Keeping these ideas—hope and success—foremost in your communication gives your child the determination to keep trying.

Steps to Solving Problems

In math, there are problem-solving steps that soon become habitual with practice. By breaking down the process into smaller pieces, you can conquer big problems. Often times, you do these steps unconsciously. You must work with your child to instill these habits. These steps break down as follows:

- **STEP 1: What is the question asking for?** This is the beginning of understanding the problem. Chapter 4 extensively covered this step. One of the strategies in Step 1, you should recall, was to restate the problem in your child's own words.
- **STEP 2: What is the given information?** Find connections between the given information and the unknown information. Chapter 4 covered some of this. Some conversation starters to give your child could include these:
    - Are there any vocabulary terms in the given information?
    - What are the similarities between the little pieces of information given?
    - What are the differences between the little pieces of information given?
- **STEP 3: Do you know or remember a similar problem?** Have you seen this before? Could you invent a new problem that is

similar? What rules from chapter 3 are imbedded in this problem? These types of questions give your child starting points to tackle the problem. Other questions are these:

- Does this problem remind you of any problems in the past?
- Can you think of any real-world examples similar to this problem?

- **STEP 4: What is the plan?** What strategies may help in the solution? Chapters 2 and 3 gave you the tools and materials needed to carry out the plan; chapters 4 and 5 develop the skills needed to use these tools. For example, if this problem reminds you of any in the past, ask these questions:
  - How did you solve the previous problem?
  - What operations or rules did you use?
  - Can you estimate an answer?
- **STEP 5: Carry out the plan.** Check each step along the way. Remember that there may be some false starts or strategies that do not result in success. Nevertheless, all attempts, whether helpful for this problem or not, are steps along the way to understanding and future success in problem solving.
- **STEP 6: Look back and forward.** Did you answer the question? Is your answer reasonable? Chapter 4 addressed these questions. Could you have done the problem differently? Can you imagine how you might use what you have learned in the future? What was difficult about this problem, and what general knowledge was learned from going through this process?

It can be helpful to have your child remember the six steps of problem solving by using a catchy acronym. The acronym could be AGSPCL, one letter for each of the steps:

- **A** represents What is being **A**sked for
- **G** represents **G**iven information
- **S** represents **S**imilar problems
- **P** represents What is the **P**lan
- **C** represents **C**arry out the plan
- **L** represents **L**ook back and forward

Adopt a silly phrase to recall the acronym, such as "All Great Students Practice Crazy Languages." Better yet, have your child come up with a funny phrase to recall the letters.

Remember, these steps are mental habits that your child can use to solve any kind of problems. Any human endeavor that results in satisfaction and self-esteem presents a challenge. By instilling this approach to learning in general, you are providing your child with life skills that extend far beyond the mathematics classroom. To compare this process and make it more real for your child, use an example from his natural interests in life. You could use examples from sports, musical instruments, puzzles, games, or electronic devices.

For example, think of how your child might approach a new video game. Remember the acronym AGSPCL, or "All Great Students Practice Crazy Languages." Step 1, **A**sk, would be getting acquainted with the game and discovering its objective. Step 2, **G**ivens, is learning how to manipulate the controls and objects to get to the game's goal. Step 3, **S**imilar, would be comparing this game to other video games that the child has played in the past and using the skills and background knowledge of gaming to tackle this new game. Steps 4 and 5, **P**lan and **C**arry out, are actually starting to play the game and attempting to succeed. Note that often, first attempts do not result in success, but every attempt is an opportunity to learn. Step 6, **L**ook back and forward, is the satisfaction of learning and doing and assessing how close the child came to conquering the game. The player may consider how she could have approached the game differently to get to the goal more efficiently. You can use an example such as this to instruct your child in the process of solving problems.

Problem-Solving Strategies

There are a number of common problem-solving strategies. When used in conjunction with the tools and materials presented in chapters 2 and 3, your child has the skills—and a better chance—at success in math.

Find Patterns

Patterns appeal to our artistic sense, and mathematics is full of patterns. Some obvious examples of patterns are lists such as {3, 6, 9, 12 . . .} (add three each time) or {10, 6, 4, 3 . . .} (divide by two and then add one each time). Patterns are also evident in a hundreds chart and in a multiplication chart. Take a multiplication or hundreds chart and encourage your child to find various patterns. Many concepts of numbers can be found in these charts. You can instill a love of patterns by pointing out patterns in everyday life, such as in mosaics, brick walkways, or bridges. Make a game out of it.

Make a Picture or Diagram

They say that a picture paints a thousand words, and this is definitely true in mathematics. A picture gives you a visual image of the situation at hand, and it is very helpful in finding connections between the given information in the problem. Drawing a picture is another way of restating the problem, and it assists in a complete understanding of the scenario. Be aware that the drawing or picture should be as simple as possible to simplify understanding. For example, use circles to represent people or a rectangle to represent a garden or swimming pool.

Work Backward

Have you or your child ever solved a maze by starting at the end and working backward to the start? If so, you have used the strategy of working backward. Some people think that this is somehow "cheating" or not authentic problem-solving. Nothing could be further from the truth. You should use any strategy in your arsenal to get the job done. Working backward is using inverse thinking and opposite operations. In fact, for some types of mathematics problems, working backward is actually the strategy of choice. Solving equations is one such mathematical exercise. Some problems give the ending result and ask you to find the starting condition. A real-life example of this is planning for retirement. You may start at the end, your target savings goal, and then work backward to calculate what you should be saving now.

Name Your Poison: Choose a Formula or Rule to Use

This strategy is used often in problem solving. Chapters 2 and 3 supplied you with some of the important vocabulary, rules, and operations needed when making a plan for solution. A firm understanding of the four operations of addition, subtraction, multiplication, and division are essential to choose the correct one. Key words can help, but the best way to choose the correct operation lies in a full understanding of the problem, Steps 1 and 2 of the plan. It is tempting to use a formula, rule, or operation and then assume the answer is correct because a rule was used to find it. This is where Step 6, **L**ook back and forward, plays an important role, making sure that the answer makes sense in the context of the problem.

Use a List or Table

Tables and lists are organizational strategies. Tables are very helpful in showing relationships between items in a problem, and organized lists are helpful in enumerating all the possibilities when combining objects. This strategy, in conjunction with looking for patterns, can be a powerful ally when tackling difficult problems.

Solve a Simpler Problem

At times, a problem is very complex, and it seems that no possible solution plan exists. A very effective strategy can be to make the problem simpler. By simplifying some of the conditions of the problem, you can then see a problem-solving method to use. This is another way to take a big problem and break it down into little pieces. Adults who design new products often use this approach. They consider a simpler problem, solve it, and then extend their solution to encompass all of the requirements of the original problem.

Guess and Check

Trial and error is inherent in many real-world problem-solving situations, and it has a place when solving math problems. It is helpful first to have a good sense of numbers so that the guesses are educated and reasonable and the checking exercise is effective.

The remainder of this chapter consists of sample problems grouped by concept. The problem-solving steps will be the guiding force in each solution. Refer back to chapter 4, if need be, for more information on Steps 1 (what is being **A**sked for), 2 (what is the **G**iven information), and 6 (**L**ook back and forward).

Number Sense and Number Operations

1. The temperature changes were recorded for one school week. On Tuesday, the temperature rose 5 degrees. On Wednesday, the temperature dropped 20 degrees. The temperature rose 6 degrees on Thursday and rose another 15 degrees on Friday for a final temperature of 13 degrees. What was the original temperature on Monday?

   There are a lot of numbers and words given in this problem, and it may, at first glance, seem intimidating. Let's go through the problem-solving process to make a plan.

   Begin with the acronym "All Great Students Practice Crazy Languages."

   A—Step 1: What is being asked for? You want to determine the original temperature on Monday. Rephrase this question to state that you want the beginning temperature before all of the rises and drops.

   G—Step 2: What is the given information? The given information is a series of rises and drops in temperatures that finally end up at the temperature of 13 degrees. Rises and drops in value involve addition and subtraction.

S—Step 3: Do you know or remember a similar problem? The answer to this question may be no, but perhaps some rules, as described in chapter 3, apply. Because you are adding and subtracting, use the rules for integer arithmetic.

P—Step 4: What is the plan? One strategy that may help with this problem is the strategy of working backward. You know the final temperature, and the question is asking for the original temperature, which is in the past. Recall that working backward involves inverse thinking and opposite operations. This will be part of the plan. You will start at 13 degrees and work backward until Monday. Drawing vertical number lines to represent the thermometers may also help your child visualize the problem.

C—Step 5: Carry out the plan. Start with the final temperature of 13 degrees and work backward by using operations opposite to the information given. Because the temperature rose 15 degrees on Friday, do the opposite operation—subtract—and take 13 – 15 to find the temperature on Thursday. Use the integer arithmetic rules explained in chapter 3 and change 13 – 15 to be 13 + –15. You are adding two numbers with different signs, so subtract and take the sign of the larger absolute value. Therefore, the temperature on Thursday was 13 + –15 = –2.

The information given states that the temperature rose 6 degrees on Thursday, so do the opposite operation and subtract to find the temperature on Wednesday. The temperature on Wednesday was –2 – 6, or –8.

The problem states that the temperature dropped 20 degrees on Wednesday, so add 20 to the temperature for Wednesday: –8 + 20 = 12 degrees on Tuesday.

Next, it is given that the temperature rose 5 degrees on Tuesday, so subtract 5 degrees to find the temperature on Monday. Monday's temperature was 12 – 5, or 7 degrees.

You can use a number line, as demonstrated in chapter 3, to remind your child of the integer addition rules.

L—Step 6: Look back and forward. Verify that you have answered the question. You now know the temperature on Monday, and that is what is being asked for. Now, try to imagine how to use a problem such as this in the future. Anytime that you are given a final result and asked to find something in the past, you can try the strategy of working backward and using inverse thinking.

It is also helpful to think of another way that the problem could be solved. This enhances your child's confidence in problem solving. In this problem, for example, you could assign a variable, *n*, to the temperature on Monday. Then, write an equation showing the changes that will finally result in 13.

$n + 5 - 20 + 6 + 15 = 13$	**The equation**
$n + 6 = 13$	**Combine like terms using integer arithmetic.**
$n + 6 - 6 = 13 - 6$	**Subtract 6 from both sides.**
$n = 7$	**The original temperature on Monday is 7 degrees.**

2. Stephanie ate $\frac{3}{5}$ of her Halloween candy the first two days. If she had a total of 75 pieces of candy, how many pieces did she eat those first two days?

There is a fraction in the problem, and often, this is enough to make a student anxious. Going through the steps will help to alleviate any concern.

Begin with the acronym "All Great Students Practice Crazy Languages."

A—Step 1: What is being asked for? The problem is asking for how many pieces she ate those first two days. Restate the question as what is $\frac{3}{5}$ of the 75 total pieces of candy?

G—Step 2: What is the given information? The problem gives the total amount of candy and the fractional part that was eaten. The fact that she ate it over two days is extra information that is not needed to solve the problem.

S—Step 3: Do you know or remember a similar problem? Think of other problems that involved fractions. A fraction, as defined in chapter 2, is used to express a ratio, in this case comparing the part eaten to the whole amount.

P—Step 4: What is the plan? The key word *of* is present in the problem, and in the table of common terms in chapter 4, it is shown that *of* means "to multiply." Therefore, $\frac{3}{5}$ of 75 is the same as $\frac{3}{5}\times 75$.

C—Step 5: Carry out the plan. Chapter 3 covered the rules for multiplying fractions. Convert 75 to the equivalent fraction $\frac{75}{1}$ and multiply $\frac{3}{5}\times\frac{75}{1}=\frac{225}{5}=45$. Stephanie ate 45 pieces of candy.

L—Step 6: Look back and forward. Is the answer reasonable? Think of the fact that $\frac{3}{5}$ is a little over one half of the candy and 45 is a little over one half of 75; the answer is reasonable. In the future, when approaching problems, you can again use your key terms and basic rules tools to arrive at solutions to problems.

3. Carlos has been saving quarters and dimes for three weeks. He counts his money and finds that he has $6.50 in quarters and dimes and a total of 44 coins. How many quarters does he have?

   Begin with the acronym "All Great Students Practice Crazy Languages."

   A—Step 1: What is being asked for? You are looking for the number of quarters that Carlos has. In other words, how many of the 44 total coins, equalling $6.50, are quarters?

   G—Step 2: What is the given information? You know the total monetary value of the coins and how many total coins. The fact that Carlos has saved for three weeks is extra, unneeded information.

   S—Step 3: Do you know or remember a similar problem? Your child may never have encountered a problem like this, but he probably knows from dealing with money that to find the value of coins, the number of a group of specific coins, such as quarters, can be multiplied by the value of one quarter, or $0.25. Take some quarters and dimes and ask your child to calculate the monetary value. To tackle this problem, consider some of the strategies from chapter 3.

   P—Step 4: What is the plan? The strategy of guess and check could be used here. Make a table that shows the relationship between the given information like the figure here:

Number of Quarters	Number of Dimes	Total Number of Coins	Monetary Value

Make educated guesses on the number of quarters and dimes. The third column in the table uses the key term *total*, which is found by adding. When guessing, ensure that the sum of the number of quarters and dimes is always 44. Calculate the monetary value by multiplying the number of quarters by $0.25 and the number of dimes by $0.10 and then adding these two values.

C—Step 5: Carry out the plan. To make an educated guess, try a number of quarters that will result in an easy calculation, such as 10. If there are 10 quarters, then there must be 34 dimes to total 44 coins. The monetary value in this case is $(10 \times 0.25) + (34 \times 0.10) = 2.50 + 3.40 = 5.90$. This monetary value is too low, so there must be more than 10 quarters. Fill in the table and make a second guess, such as 15 quarters. If there are 15 quarters, then there must be $44 - 15 = 29$ dimes. This monetary value is $(15 \times 0.25) + (29 \times 0.10) = 3.75 + 2.90 = 6.65$. This guess is too high in value, but it is very close to the correct value. Fill in the table. Try 14 quarters for the third guess. If there are 14 quarters, then there are 30 dimes. The money value is $(14 \times 0.25) + (30 \times 0.10) = 3.50 + 3.00 = 6.50$. This is correct!

Number of Quarters	Number of Dimes	Total Number of Coins	Monetary Value
10	34	44	$5.90
15	29	44	$6.65
14	30	44	$6.50

L—Step 6: Look back and forward. Verify that you answered the question. There are 14 quarters and 30 dimes, but the problem is only asking for the number of quarters; the correct answer is 14. Using guess and check and a table are effective strategies when there is a relationship between two items.

4. At Lincoln Elementary School, there are 5 girls for every 4 boys enrolled. If there are 250 girls enrolled, how many boys attend the school?

Begin with the acronym "All Great Students Practice Crazy Languages."

A—Step 1: What is being asked for? The problem is asking how many boys attend the school. Put the question in your own words, such as find the number of boys when you know the number of girls at the school.

G—Step 2: What is the given information? The first sentence of the problem describes a comparison of girls to boys, which is a ratio. The problem also states that there are 250 girls in the school.

S—Step 3: Do you know or remember a similar problem? A comparison is a ratio. Have you seen a problem like this before? Chapter 3 introduced you to problems like this in the section on ratio and proportion! Chapter 3 also reviewed the rules for cross-multiplication with proportions.

P—Step 4: What is the plan? Set up a proportion comparing girls to boys. Use a variable, *n*, for the unknown number of boys in the school. Cross-multiply to find the value of *n*.

C—Step 5: Carry out the plan. Set up the proportion:

$\frac{5}{4} = \frac{250}{n}$	**The proportion comparing girls to boys**
$5n = 1{,}000$	**Cross-multiply.**
$\frac{5n}{5} = \frac{1{,}000}{5}$	**Divide both sides by 5 to isolate the variable *n*.**
$n = 200$	**The number of boys in the school**

L—Step 6: Look back and forward. Did you answer the question? Yes, you have determined the number of boys in the school. You can be reasonably sure that this is correct because the number of boys is less than the number of girls. If you had set up the proportion incorrectly, you would have found that the number of boys was larger than the number of girls, and that would be a red flag to go back and recheck.

It would be instructive and build confidence if you could think of another way to solve this problem. Another way to approach this problem would be to divide the number of girls by 5 to get $250 \div 5 = 50$. Because the number of girls in the school is $50 \times 5 = 250$, the number of boys is $50 \times 4 = 200$ boys.

Measurement

1. A table is 225 centimeters in length. How many meters long is the table?

   1 meter = 100 centimeters

   Begin with the acronym "All Great Students Practice Crazy Languages."

   A—Step 1: What is being asked for? The problem is asking how many meters long the table is. Another way to state the question is to convert from centimeters to meters.

   G—Step 2: What is the given information? The table is 225 centimeters long, and 1 meter is equal to 100 centimeters. Show your child 100 centimeters on a meter stick.

   S—Step 3: Do you know or remember a similar problem? You are being asked to do a conversion. Have you seen a similar problem?

Do you remember the conversion problem outlined in chapter 4 where you were asked to convert quarts to gallons? It is not exactly the same problem, but the methods used in that example may help in this problem.

P—Step 4: What is the plan? The question is looking for the number of meters and gives you the number of centimeters. Because 1 meter is equivalent to 100 centimeters, the number of meters will be a smaller value. Divide the number of centimeters, 225, by 100 to find the number of meters.

C—Step 5: Carry out the plan. Divide: 225 ÷ 100 = 2.25 meters.

L—Step 6: Look back and forward. Did you answer the question? Yes, you found the number of meters. This answer makes sense because it is a smaller number than the number of centimeters.

Can you think of another way to solve this problem? Well, this is a comparison between centimeters and meters. A comparison could be set up as a proportion, as we did in Example 4 under "Number Sense and Number Operations." Set up the proportion:

$\frac{100}{1} = \frac{225}{n}$	**The proportion comparing centimeters to meters**
$100n = 225$	**Cross-multiply.**
$\frac{100n}{100} = \frac{225}{100}$	**Divide both sides by 100 to isolate the variable *n*.**
$n = 2.25$	**The number of meters**

Considering alternate methods of solution is a valuable process that strengthens your child's understanding of the connections in mathematics. Now, your child has two strategies to use to tackle conversion problems!

2. What is the area of a rectangle on a coordinate plane with vertices at $A\,(1, 2)$, $B\,(7, 2)$, $C\,(7, 5)$ and $D\,(1, 5)$?

   Begin with the acronym "All Great Students Practice Crazy Languages."

   A—Step 1: What is being asked for? You need to find the area of a rectangle.

   G—Step 2: What is the given information? The given information is that the figure is a rectangle and it has vertices that are ordered pairs on a coordinate plane.

   S—Step 3: Do you know or remember a similar problem? Chapter 3 discussed the coordinate plane with instruction on how to graph points. In addition, your child has most likely encountered problems involved with finding the area of a rectangle. If you have any floors in your home made up of square tiles, you can help your child with the concept that area is the number of square units that it takes to cover a figure. Alternatively, you can use something such as square tiles from a board game to demonstrate area. Show how the area (the number of square tiles) relates to multiplication (multiplying the length times the width of a rectangular figure). The formula for the area of a rectangle will be needed to solve this problem.

   P—Step 4: What is the plan? The first strategy is to draw a picture of the situation to help make the understanding clear. Another strategy is to rephrase the question. Following the steps outlined in chapter 3, graph the rectangle on the coordinate plane:

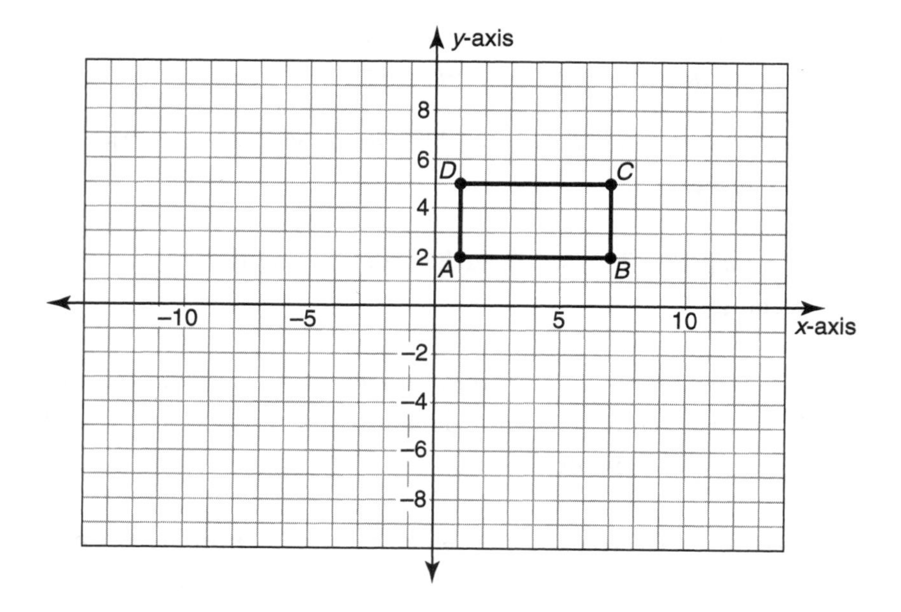

Now that there is a picture of the problem, you can get a better idea of how to find the area. Use the formula for the area of a rectangle, $A = bh$, where b is the base length and h is the height. To find these length values, count the number of squares from the diagram and then calculate the area.

C—Step 5: Carry out the plan. Refer to the diagram and count the number of squares in the base, which is 6 units long. By counting again, the height is 3 units in length. The area is, therefore, $A = 6 \times 3 = 18$ square units.

L—Step 6: Look back and forward. Did you answer the question? Yes, you found the area of the rectangle. This answer makes sense because you can actually count the number of square units within the rectangle and verify that it is 18. In looking forward, help your child to realize that drawing a picture of the situation was a fundamental strategy to arrive at the solution here. This is almost always the case when the problem involves geometry.

3. This right triangle shows the length of two sides.

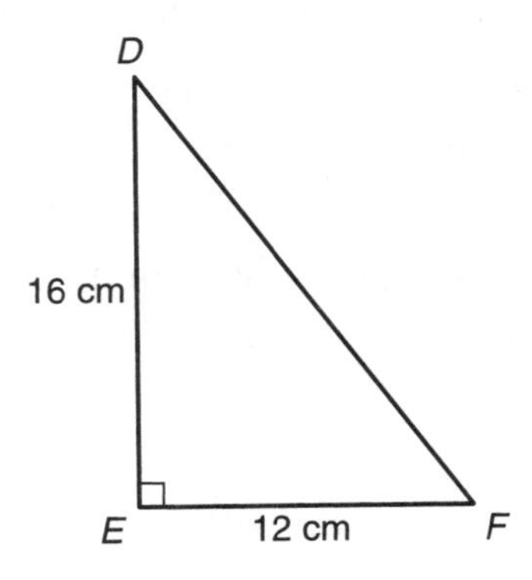

What is the length of side *DF*?

Begin with the acronym "All Great Students Practice Crazy Languages."

A—Step 1: What is being asked for? The length of side *DF*.

G—Step 2: What is the given information? This problem is a split-graphic question, as described in chapter 4. There is information above the graphic, on the graphic, and below the graphic. It is given that the triangle is a right triangle, with the two shorter sides of length 12 cm and 16 cm.

S—Step 3: Do you know or remember a similar problem? This is a measurement problem, and you are asked to find a length. In similar measurement problems that involve geometry, a formula is often used to find a solution. In chapter 3, you saw a similar problem with a right triangle in the "Using Formulas" section. It is not exactly the same problem, but the solution to that problem will most likely help in solving this one.

P—Step 4: What is the plan? The strategy to use is a formula. Try to think of a formula that will solve for the unknown length. You are told that this is a right triangle. The formula to use is the Pythagorean theorem, as reviewed in chapter 3. The formula is $a^2 + b^2 = c^2$, where a and b are the shorter sides of the triangle, called the legs, and c is the longest side, known as the hypotenuse. From the drawing shown, the longest side, c, is the length of *DF*. You can use straws or strands of dry spaghetti to make right triangles of different sizes. Demonstrate through comparison that the longest side is always across from the right angle.

C—Step 5: Carry out the plan.

$a^2 + b^2 = c^2$	**The Pythagorean theorem**
$12^2 + 16^2 = c^2$	**Substitute the given values.**
$144 + 256 = c^2$	**Evaluate the exponents.**
$400 = c^2$	**Combine like terms.**
$\sqrt{400} = \sqrt{c^2}$	**Undo the exponent of 2 by taking the square root of both sides.**
$c = 20$ cm	**The measure of the missing side with units included**

L—Step 6: Look back and forward. Did you answer the question? Yes, you found the length of *DF*, which is 20 cm. This answer makes sense because 20 is the longest length and *DF* is the longest side of the triangle. Look forward and remember that in the future, when a problem involves measurement and geometry, a formula is most often needed to solve the problem.

4. Find the length of a diagonal of a rectangle with sides of 5 inches and 12 inches.

   Begin with the acronym "All Great Students Practice Crazy Languages."

A—Step 1: What is being asked for? You need to find length of the diagonal of a rectangle.

G—Step 2: What is the given information? The given information is a rectangle with side lengths of 5 and 12 inches. The problem refers to a diagonal, which was defined in chapter 2.

S—Step 3: Do you know or remember a similar problem? This problem seems unfamiliar. Have your child think back to other problems. Remind her that this is a geometry measurement problem. In the past, drawing a picture has been helpful, as well as using a formula.

P—Step 4: What is the plan? The first strategy is to draw a picture of the situation. Remind your child that this is a way to make the understanding clear and will help rephrase the question.

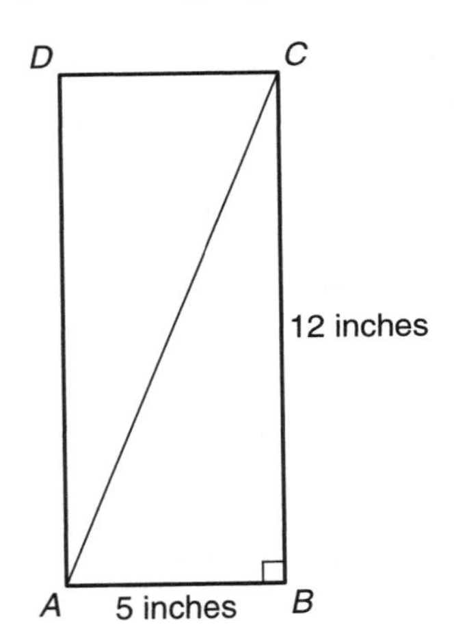

Now that there is a picture of the problem, you can get a better idea of how to find the length of the diagonal. At first glance, this problem may seem impossible to solve. There is no set formula for finding the length of a diagonal. Have your child look again at the picture. Could any smaller pieces lead to a plan? Are any smaller figures present besides a rectangle? By taking another look at the diagram, you may notice that two right triangles are imbedded in the rectangle. Now, the problem is looking similar to the previous problem in this section—a right triangle problem in which you are looking for the length of the longest side of the triangle! This stage of discovering a right triangle in the diagram may not come naturally at first, but once developed, this skill is crucial in being a first-class math problem solver.

C—Step 5: Carry out the plan. You know from the picture that this problem involves a right triangle with the two shorter side lengths of 5 and 12 inches and that you are going to use the Pythagorean theorem to find the longest side, which is also the diagonal of the rectangle.

$a^2 + b^2 = c^2$	**The Pythagorean theorem**
$5^2 + 12^2 = c^2$	**Substitute the given values.**
$25 + 144 = c^2$	**Evaluate the exponents.**
$169 = c^2$	**Combine like terms.**
$\sqrt{169} = \sqrt{c^2}$	**Undo the exponent of 2 by taking the square root of both sides.**
$c = 13$ inches	**The measure of the diagonal with units included.**

L—Step 6: Look back and forward. Did you answer the question? Yes, you found the length of the diagonal. In looking forward, help your child to realize that drawing a picture of the situation was again a fundamental strategy to arrive at the solution. In addition, it was necessary to look at the geometric figure and find smaller geometric figures within it. This skill will be used often in high school geometry work. By discussing the problem in this way, you provide your child with closure on the problem and give her a chance to anticipate what skills were used and how these same skills may be needed in future math work.

Geometry

1. The measurements of two of the angles in a triangle are 30° and 45°. What type of angle is the third angle of this triangle?

   Begin with the acronym "All Great Students Practice Crazy Languages."

   A—Step 1: What is being asked for? The problem is asking for the type of angle of the third angle of the triangle. It is not asking for the angle measurement but for the classification of the missing angle measurement.

   G—Step 2: What is the given information? You are given a triangle with two angle measures of 30° and 45°.

   S—Step 3: Do you know or remember a similar problem? Have your child try to recall a time that he has encountered a problem with the angles in a triangle. Have you seen a similar problem? In chapter 4, under the section "Does My Answer Make Sense?," Example 3 dealt with the angles in a triangle. In addition, chapter 2 defined types of angles: an acute angle measures less than 90°, a right angle measures exactly 90°, and an obtuse angle measures greater than 90° and less than 180°.

   P—Step 4: What is the plan? To classify an angle, you need the measurement of the angle. Use the fact that the sum of the measures of the angles in a triangle is equal to 180° to find the missing angle. You can verify this with your child by creating several triangles, using straws or dry spaghetti, and measuring the three angles. They will always sum to 180°. Then determine the angle type as acute, right, or obtuse by using the definitions provided in Step 3.

C—Step 5: Carry out the plan. Use a variable, *n*, to represent the missing angle measurement. Set up the situation mathematically and solve the equation:

$n + 30 + 45 = 180$	**The equation to solve**
$n + 75 = 180$	**Combine like terms.**
$n + 75 - 75 = 180 - 75$	**Subtract 75 from both sides.**
$n = 105$	**This is the measure of the missing angle.**

L—Step 6: Look back and forward. Did you answer the question? You found the measure of the missing angle, but that is not what is being asked. Finish the problem by naming the type of angle. Because the measure is greater than 90° and less than 180°, the angle is an obtuse angle.

2. Two angles in a quadrilateral have measurements of 95° and 75°. The other two angles have the same measurements as each other. What is the measurement of each of the other angles?

   Begin with the acronym "All Great Students Practice Crazy Languages."

   A—Step 1: What is being asked for? The problem is asking for the measurement of each of the two missing angles in a quadrilateral.

   G—Step 2: What is the given information? You are given a four-sided figure with two angle measurements of 95° and 75°.

   S—Step 3: Do you know or remember a similar problem? Have you seen a similar problem? Yes, in fact you have—in the previous example! The last example dealt with the angles in a triangle, and this problem deals with the angles in a quadrilateral. The solution process is similar.

P—Step 4: What is the plan? To find the missing angle measurements, you need to know the sum of the measurements of the angles in a quadrilateral. If this is not known, have your child reference her math text book or other resource to discover that the sum of the angles in a quadrilateral is equal to 360°. Two angles are missing, but they have equal measures. You can subtract the sum of the measurements of the two angles given from 360 and then divide by 2.

C—Step 5: Carry out the plan. $360 - (95 + 75) = 360 - 170 = 190$. The two missing angles together measure 190°. Because they are equal in measure, each of the angles measures $190 \div 2 = 95$ degrees.

L—Step 6: Look back and forward. Did you answer the question? You found the measure of each of the missing angles. Test this answer by verifying that $75 + 95 + 95 + 95 = 360$.

It can be helpful to have your child consider an alternate way to solve the problem. Following the method in the previous example, you could use a variable, such as x, to represent the missing angle measurement. Set up the situation mathematically as $x + x + 75 + 95 = 360$. Solve this equation:

$$x + x + 75 + 95 = 360$$
The equation to solve

$$2x + 170 = 360$$
Combine like terms.

$$2x + 170 - 170 = 360 - 170$$
Subtract 170 from both sides.

$$2x = 190$$
Combine like terms.

$$\frac{2x}{2} = \frac{190}{2}$$
Divide both sides by 2.

$$x = 95$$
The measurement of each of the missing angles

While this process is more involved, the extra effort is not without benefit. You have helped your child see the important connection between algebra and problem solving.

3. The rectangles shown in the following illustration are similar.

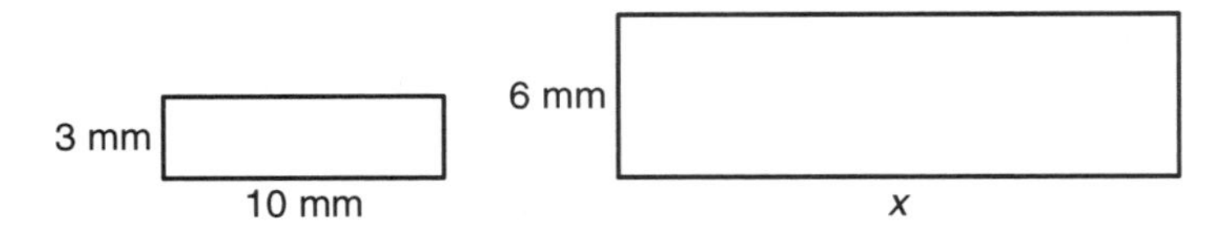

What is the value of x ?

Begin with the acronym "All Great Students Practice Crazy Languages."

A—Step 1: What is being asked for? The problem is asking for the value of the variable x. To restate the question, find the missing side length of the larger of two similar rectangles.

G—Step 2: What is the given information? There are two similar rectangles. The smaller rectangle has dimensions of 3 mm and 10 mm. The larger rectangle has a shorter side length of 6 mm, which corresponds to the length of 3 mm in the smaller rectangle.

S—Step 3: Do you know or remember a similar problem? Have you seen a similar problem? Perhaps your child has never encountered a problem like this one before. Go back to chapter 2 and find the definition of the word *similar*. Similar figures have sides that are in proportion.

P—Step 4: What is the plan? Because the sides of the two rectangles are in proportion, the strategy will be to use the rules for solving proportions by cross-multiplication, as outlined in chapter 3.

C—Step 5: Carry out the plan. Set up the proportion, being careful to keep the corresponding sides in the correct ratio. You may set up the proportion as $\frac{big}{little} = \frac{big}{little}$.

$\frac{6}{3} = \frac{x}{10}$	**The proportion**
$3x = 60$	**Cross-multiply.**
$\frac{3x}{3} = \frac{60}{3}$	**Divide both sides by 3.**
$x = 20$	**The missing side is 20 mm.**

L—Step 6: Look back and forward. Did you answer the question? You found the length of the missing side of the rectangle. Does this answer make sense? Well, consider another way to solve the problem and see if the answers agree. In chapter 2's "Synonyms" section, an alternate method of making sense of similar figures was described. Look at the relationship of the shorter sides of each of the rectangles. The larger rectangle's shortest side is twice the measure of the smaller rectangle's shortest side. Therefore, the longer side of the larger rectangle should also be twice the measure of the longer side of the smaller rectangle. Because $10 \times 2 = 20$ mm, the answer agrees with the original conclusion.

Algebra

1. Evaluate the expression when $x = 5$: $(x + 4)^2 \div (1 + 8 \div 4)$.

   Begin with the acronym "All Great Students Practice Crazy Languages."

   A—Step 1: What is being asked for? You are asked to evaluate an expression. Restate this problem, as you must substitute a given value for *x*, and then simplify.

G—Step 2: What is the given information? It is given that the variable, x, is equal to 5 and the expression to be simplified is $(x + 4)^2 \div (1 + 8 \div 4)$.

S—Step 3: Do you know or remember a similar problem? In an expression with multiple numbers and operations, you must use the order of operations to simplify.

P—Step 4: What is the plan? Use the rule for the order of operations, as explained in chapter 3. Substitute the value of 5 everywhere there is the variable x and then apply the order of operations.

C—Step 5: Carry out the plan.

$(x + 4)^2 \div (1 + 8 \div 4)$	**The given expression**
$(5 + 4)^2 \div (1 + 8 \div 4)$	**Substitute the value of 5 for *x*.**
$9^2 \div (1 + 2)$	**Evaluate in parentheses first; in the second set of parentheses, evaluate division before addition.**
$9^2 \div 3$	**Finish evaluating within the parentheses.**
$81 \div 3$	**Evaluate exponents.**
27	**Divide to simplify the expression.**

L—Step 6: Look back and forward. Many times, to solve a problem, you will need to remember similar problems and any rules, formulas, or definitions that may assist in your plan to arrive at the answer. This problem required substitution and the correct order of operations to find the solution, both of which are common processes for most students at the middle or high school level. You may want to review these, using chapter 3 as a reference, if your child experienced difficulty with this or a similar problem.

2. Solve for x: $x + 12 = 74$.

   Begin with the acronym "All Great Students Practice Crazy Languages."

   A—Step 1: What is being asked for? You are asked to solve an equation. This could be reworded as "isolate the variable x to find its value."

   G—Step 2: What is the given information? An equation is given.

   S—Step 3: Do you know or remember a similar problem? Your child may have solved similar equations in the past, in which case he knows how to proceed. If not, have your child go back to chapter 3 and review the techniques for solving equations.

   P—Step 4: What is the plan? To isolate the variable, you use the strategy of working backward by applying opposite operations. This procedure is done until the variable is alone on one side of the equation.

   C—Step 5: Carry out the plan.

$x + 12 = 74$	**The given equation**
$x + 12 - 12 = 74 - 12$	**The opposite operation to addition is subtraction. Subtract 12 from each side.**
$x = 62$	**Isolate the variable; solve the equation.**

   L—Step 6: Look back and forward. Remember that it is a good practice to check the answer to ensure that it is correct. A check is done by substituting the value arrived at for x in the original equation and making sure you get a true statement.

$x + 12 = 74$	**The original equation**
$62 + 12 = 74$	**Substitute the value of 62 for *x*.**
$74 = 74$	**This is a true statement; the answer checks.**

If your child understands that the check ensures he is on the right track, solving equations can be a satisfying experience.

3. A book club charges a one-time fee of $20 to join and then $3 for each book bought. The following equation shows the relationship between the number of books purchased and the total fees, including the one-time fee, in dollars.

$$3x + 20 = 56$$

How many books were purchased if the total fees were $56?

Begin with the acronym "All Great Students Practice Crazy Languages."

A—Step 1: What is being asked for? The problem requires you to find how many books were purchased. If you notice that an equation is given, you can restate the problem as asking you to solve the given equation for x.

G—Step 2: What is the given information? A description of the fees is given, as well as the equation needed to solve the problem.

S—Step 3: Do you know or remember a similar problem? While this problem differs from the previous problem in that it contains a lot of words, an equation is given that needs to be solved, just as in the last example.

P—Step 4: What is the plan? To isolate the variable, you use the strategy of working backward by applying opposite operations. This procedure is done until the variable is alone on one side of the equation. Remember that in a two-step equation, such as this one, you undo the addition before you undo the multiplication, which is the inverse process to the order of operations.

C—Step 5: Carry out the plan.

$3x + 20 = 56$	**The given equation**
$3x + 20 - 20 = 56 - 20$	**Subtract 20 from each side.**
$3x = 36$	**Combine like terms.**
$\frac{3x}{3} = \frac{36}{3}$	**Divide both sides by 3.**
$x = 12$	**Isolate the variable.**

The total number of books purchased was 12.

L—Step 6: Look back and forward. Remember that it is a good practice to check the answer to ensure that it is correct. In a word problem, is it best to go back to the words to double-check that the purchase of 12 books is correct. If 12 books were purchased, the fees would be the one-time fee of \$20 plus the fee for the books, $12 \times 3 = \$36$. Because $20 + 36 = 56$, the check works, and the answer is correct.

4. If the area of the triangle shown is 60 cm^2, what is the height?

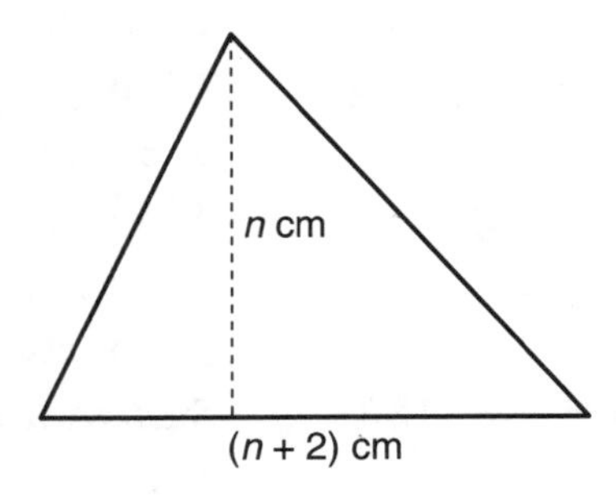

Begin with the acronym "All Great Students Practice Crazy Languages."

A—Step 1: What is being asked for? The problem is asking for the height of the triangle.

G—Step 2: What is the given information? The area of the triangle is given as 60 cm^2, the base is expressed as the variable expression $n + 2$, and the height is expressed as the variable n.

S—Step 3: Do you know or remember a similar problem? This is a problem dealing with geometry and measurement, so the formula for the area of a triangle will be needed; that is, $A = \frac{1}{2}bh$, where b is the base of the triangle and h is the height.

P—Step 5: What is the plan? This problem has variables instead of just numbers. You could use the strategy of making the problem simpler. For example, if the base and height were just numbers, you would substitute these numbers into the area formula. This is exactly what you should do in this problem but, instead, substitute in the variable expressions, just as you would if they were simply numbers.

C—Step 6: Carry out the plan.

$A = \frac{1}{2}bh$	**Start with the formula for the area of a triangle.**
$60 = \frac{1}{2}n(n+2)$	**Substitute in the given information.**
$60 \times 2 = \frac{1}{2} \times 2 \times n(n+2)$	**Multiply both sides by 2 to clear the fraction.**
$120 = n(n + 2)$	**The simplified equation**
$120 = n^2 + 2n$	**Apply the distributive property.**
$0 = n^2 + 2n - 120$	**Subtract 120 from both sides.**
$0 = (n + 12)(n - 10)$	**Factor the trinomial.**
$n = -12$ or $n = 10$	**The two solutions**

The negative solution, $n = -12$, must be rejected because you cannot have a negative length. Therefore, the height of the triangle, n, is 10.

L—Step 6: Look back and forward. The check of this solution is straightforward. If the height, n, is 10, then the base is $n + 2 = 10 + 2 = 12$. Use these values in the formula for the area of a triangle to verify that the area is 60 cm^2, as given in the information of the problem: $A = \frac{1}{2}bh = \frac{1}{2} \times 12 \times 10 = 60$ cm^2.

An alternate way to solve this problem without algebra is to use the guess and check strategy. You are told that the area is 60 cm^2 and that the base is two more than the height. You could make a table of base and height guesses, where the base is two more than the height each time, and calculate areas until the target area is reached. By emphasizing the idea that there are multiple ways to solve the problem, you build your child's supply of strategies to use to be an effective problem solver.

Probability and Statistics

1. How many lunch selections, consisting of one sandwich, one soup, and one drink, are possible if lunch can be chosen from the following table?

Lunch Selections

Sandwich Choices	Soup Choices	Drink Choices
Ham (H)	Vegetable (V)	Milk (M)
Turkey (T)	Chicken Noodle (C)	Juice (J)
Peanut Butter (P)		

Begin with the acronym "All Great Students Practice Crazy Languages."

A—Step 1: What is being asked for? You are asked to calculate the number of lunch selections possible. This is the total number of combinations consisting of one sandwich, one soup, and one drink.

G—Step 2: What is the given information? A lunch selection is one sandwich choice, one soup choice, and one drink choice. From the information in the table, there are three choices of sandwiches, two choices of soups, and two choices of drink.

S—Step 3: Do you know or remember a similar problem? An identical type of problem, dealing with ice cream cones, was introduced in the last chapter. Use a similar approach to solving this problem.

P—Step 4: What is the plan? Use the strategies of making a diagram, called a tree diagram, and then use the diagram to make an organized list of the possibilities.

C—Step 5: Carry out the plan. Make a tree diagram of the situation.

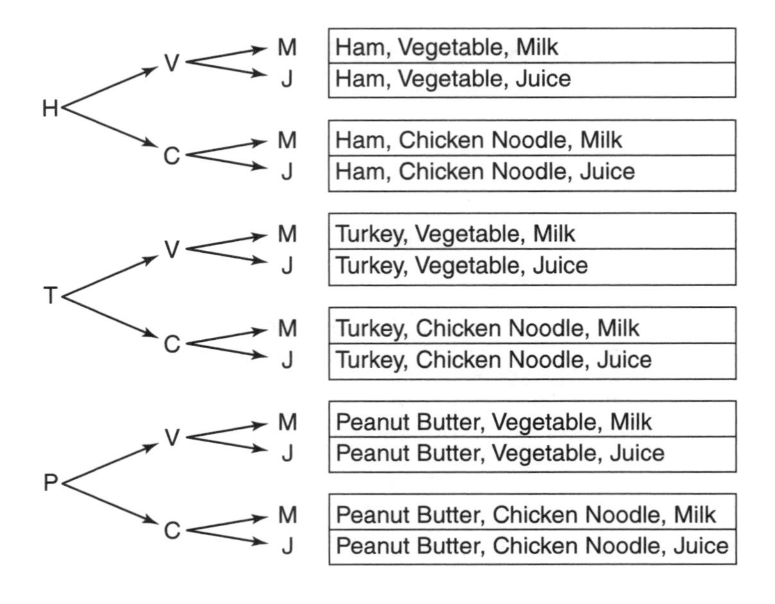

The tree diagram, as shown, is a way to make an organized list of all of the possibilities. There are 12 different possibilities for lunch selections.

L—Step 6: Look back and forward. There are 12 different lunch selection possibilities. An alternate way to solve the problem is to use the counting principle as described in chapter 4 with the ice cream cone example. The number of different lunch selections is $3 \times 2 \times 2 =$ 12 selections. The tree diagram illustrates this principle because there are two groups of two and one group of three, which means you can multiply to solve.

2. Using the spinner shown, what is the probability of spinning a three and then spinning a three again?

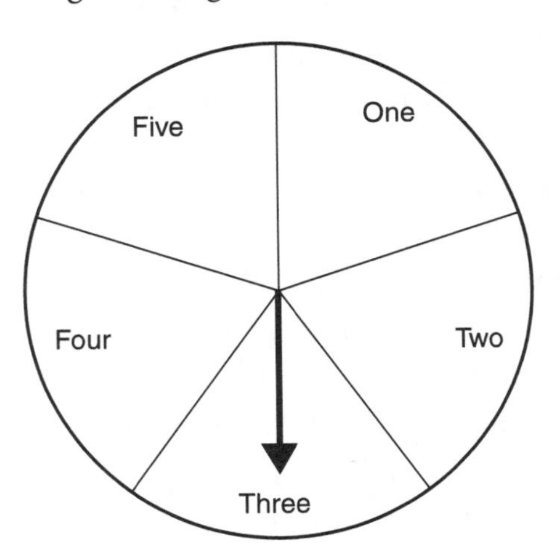

Begin with the acronym "All Great Students Practice Crazy Languages."

A—Step 1: What is being asked for? The question is asking for the probability of spinning a three and then a three again.

G—Step 2: What is the given information? The given information is a spinner with five equal sections, one of which is labeled "Three."

S—Step 3: Do you know or remember a similar problem? Probability was defined in the vocabulary section of chapter 2, and an example was given in chapter 4.

P—Step 4: What is the plan? Probability is defined as the ratio of the $\frac{\text{number of ways an event can occur}}{\text{total number of possibilities}}$. The total number of possibilities can be found by the counting principle, previously explained. There are $5 \times 5 = 25$ possibilities when you spin the spinner two times, because there are five choices each time you spin.

C—Step 5: Carry out the plan. There are 25 different possibilities when you spin the spinner two times, and only one of these possibilities is the event of spinning a three and then a three again. The probability is therefore $\frac{1}{25}$. To help understand probability, create a spinner and have your child spin it numerous times, just to see that she does not very often spin a three and then a three again.

L—Step 6: Look back and forward. Does this answer make sense? Sometimes, it is hard to be sure when you are dealing with the probability of multiple events. However, you can double-check this result by solving the problem in an alternate way. The probability of one event, E_1, and then another, E_2, can be found by multiplying the probability of the first event by the probability of the second event. This is written as $P(E_1, E_2) = P(E_1) \times P(E_2)$. The probability of spinning a three is $\frac{1}{5}$. Therefore, $P(\text{three, three}) = \frac{1}{5} \times \frac{1}{5} = \frac{1}{25}$.

3. The following cards are in a bag. You randomly pick a card without looking. You do not replace this card, and then you randomly pick another card. What is the probability of picking an *A*, and then picking a *P* ?

Begin with the acronym "All Great Students Practice Crazy Languages."

A—Step 1: What is being asked for? The question asks you to find the probability of choosing an A, not replacing it, and then choosing a P. Probability tells the likelihood of this event happening.

G—Step 2: What is the given information? The information gives you nine cards with letters printed on them. One card is chosen, not put back in the bag, and then another card is chosen.

S—Step 3: Do you know or remember a similar problem? The previous problem also dealt with probability. Some parts of that solution may help in this problem.

P—Step 4: What is the plan? Probability is defined as the ratio of the $\frac{\text{number of ways an event can occur}}{\text{total number of possibilities}}$. When two events occur, like choosing a card two times, the probability of one event, E_1, and then another, E_2, can be found by multiplying the probability of the first event by the probability of the second event. This is written as $P(E_1, E_2) = P(E_1) \times P(E_2)$. Using this rule, find the probability of each separate event and then multiply them together.

C—Step 5: Carry out the plan. The probability of choosing an A is $\frac{1}{9}$, because there is one A out of a total of 9 cards. Now that the A has been chosen and not put back, there are 8 cards left. Therefore, the probability of choosing a P at this time is $\frac{2}{8} = \frac{1}{4}$ because there are 2 P's out of a total of 8 cards. The probability is, therefore, $\frac{1}{9} \times \frac{1}{4} = \frac{1}{36}$. To help you understand, you could actually use index cards to create these letter cards. Place them in a bag and have your child act out this problem.

L—Step 6: Look back and forward. Another way to think about this problem is to realize that there are $9 \times 8 = 72$ total possibilities for picking two cards, when you do not replace the first one. Out of these 72 possibilities, two events could result in an *A* and then a *P*. There are two events, because there are actually two separate *P* cards in the bag: $\frac{2}{72}=\frac{1}{36}$, and this agrees with our answer.

4. The following data shows the amount of time that Lara spent at the gym each day for one week.

Sunday	Monday	Tuesday	Wednesday	Thursday	Friday	Saturday
80 min	60 min	75 min	0 min	45 min	30 min	60 min

What is the mean time in minutes that she spent each day?

Begin with the acronym "All Great Students Practice Crazy Languages."

A—Step 1: What is being asked for? The problem is asking you to find the mean time in minutes that she spent at the gym each day. This can be restated to say the average time spent each day at the gym.

G—Step 2: What is the given information? The problem supplies the amount of time spent each day for a seven-day period.

S—Step 3: Do you know or remember a similar problem? Your child may or may not have calculated the mean before. This concept was introduced in the vocabulary section of chapter 2, as well as the "Synonyms" section of the same chapter.

P—Step 4: What is the plan? To find the mean, you add up all of the data and then divide by the number of data items in the set. In this problem, there are seven pieces of data.

C—Step 5: Carry out the plan. Add all of the data items: 80 + 60 + 75 + 0 + 45 + 30 + 60 = 350. Now, divide this sum by 7, the number of days: 350 ÷ 7 = 50 minutes.

L—Step 6: Look back and forward. Is this answer reasonable? Whenever you compute a mean, make sure that you go back and look at the data to be sure that the mean you calculated makes sense. For the data given in the table, some of the values are higher than 50, and some are lower than 50. Because there is also a data value of 0, the mean is smaller than most of the data. This answer is reasonable. Often times, students forget to divide by the number of data items when calculating the mean. If they go back and compare their value to the data set, they quickly realize that the mean does not make sense.

5. Sahiba has the goal of achieving a mean grade average of 94 for the first semester. She has four grades so far and one more upcoming test grade for the semester. Given her grades so far, what grade must she earn on the upcoming test to reach her goal?

Grade #1	Grade #2	Grade #3	Grade #4	Grade #5
96	88	98	92	???

Begin with the acronym "All Great Students Practice Crazy Languages."

A—Step 1: What is being asked for? The question is asking what grade she must earn on the upcoming test to have a mean grade average of 94.

G—Step 2: What is the given information? The problem gives the first four grades and the target mean grade goal of 94.

S—Step 3: Do you know or remember a similar problem? The previous problem is similar, in that it uses the concept of mean. This problem is a little different, however, because it has a different unknown.

P—Step 4: What is the plan? Because the mean is given and you are looking for a piece of data, use the strategy of working backward. First, multiply the target mean number by 5, which is the number of data items that includes the fifth grade. Then subtract all of the given grades to find the missing grade. This is the opposite operation of computing the mean itself.

C—Step 5: Carry out the plan. The target mean grade is 94, so multiply 94 by 5, the number of data items: $94 \times 5 = 470$. Now, subtract the four given grades: $470 - 96 - 88 - 98 - 92 = 96$. Sahiba needs to earn a 96 on the fifth test to have a mean grade of 94 for the semester.

L—Step 6: Look back and forward. Is this answer reasonable? Well, you can easily check if it is correct by computing the mean with the five data items with the fifth grade equal to 96. Add up the data items: $96 + 88 + 98 + 92 + 96 = 470$. Now, divide this sum by 5 to get $470 \div 5 = 94$. Working backward is a powerful strategy to use because you can always check your work by working forward with the answer you arrive at.

What Have We Learned About Problem Solving?

This chapter has given you, the parent, some insight into the problem-solving process. It has made conscious for you the steps that people often do unconsciously to arrive at a plan for handling math problems. If you go over these steps with your child verbally and work through these examples, you will give your child the skills necessary to go about the job of solving math problems. Working through these steps out loud begins to assimilate them into your child's habits of mind, and the steps will become more natural and automatic. Remember the acronym AGSPCL, or "All Great Students Practice Crazy Languages."

This chapter also supplied you with some common problem-solving strategies. They will help with creating a plan for solution. These strategies, in addition to the information in the previous chapters in this book, give your child a problem-solving tool box of ideas to pull from when faced with difficult math problems.

Be creative and look for opportunities to act out problems. This will help to make the problems more real to your child.

Math problems are really just a collection of little pieces. These little pieces are easier to deal with, and when you systematically attack problems in these little pieces, you realize that you are solving big problems!

Chapter 6

Study Strategies

One of the most important aspects of mastering new concepts is developing good study habits and strategies. Often, students are told to "study" for a test, when they really have no idea what this means or where to start. In this chapter, specific tips for studying mathematics are discussed, in turn giving you ways you can help your child establish these lifelong good habits of learning. Strategies for establishing a set schedule and study location, using a math textbook to its potential, and approaching homework and tests are explained in detail. For further information, see the "What kind of learner is your child?" quiz in the appendix, which suggests strategies specific to your child's learning style.

Establish a Set Schedule

One of the first things to do is to establish a set schedule for studying. This is time allotted for math homework and a review of concepts. Math is a cumulative subject, where each new concept builds on ideas and procedures previously studied. It is important that each of these topics be mastered before advancing to more complicated ideas.

In many math classrooms, there is a math assignment almost every night. These assignments allow the student to practice concepts and reinforce skills worked on in the classroom setting. Because of this, the time set aside for math homework and practice should be consistent throughout the week, and ideally, at the same time each day. This way, the importance of math homework and practicing the strategies becomes apparent to your child and becomes part of his daily activities. Even if there is not an assignment each day, a few minutes can be devoted to reinforcing basic skills and concepts as your child continues the study of mathematics.

Find a Study Spot

Finding a location to study that is quiet and free of distractions is as important as finding the time to study. Choose a spot that the child can use each day. The studying and concentration take place in this designated area.

It would be to your child's advantage to keep this area free of toys and other distractions. The television and music should be shut off. Make these activities the reward for after studying. On the floor, in front of the television, is not the correct place for study. It is good to have a clutter-free area for your child to work. An ideal spot would be a desk in your child's bedroom. If your assistance is needed, the child can work at the dining room or kitchen table.

Using a Math Textbook

Math textbooks present a challenge to both students and parents. The technical reading required to interpret the ideas and examples in each section of the book can be difficult to understand.

Some important activities can help your child understand the text scattered throughout the book like a scavenger hunt. With your child, take a walk through the textbook. During this process, have your child identify such things as the table of contents, index, and glossary and practice looking through the chapters to see what's included. Talk about the fact that each chapter has a main subject, or idea, and each section pertains to this concept.

For the first activity, with your child find the glossary in the back of the book. Discuss the fact that definitions of concepts, and sometimes examples, are included in a glossary and that all of the terms are in alphabetical order. Have your child practice looking up a word and read the definition together. If an example is given, show how this can help to clarify the definition.

Next, find the index of the book. The index is a listing of each of the terms and concepts presented in the book, along with the page numbers on which the term is found. Use the index with your child to find, perhaps, a concept currently being studied in class or a topic with which she has had particular difficulty. Knowing about the index and how to use it adds another tool to your child's mathematical toolbox.

Walk through a section in the first unit. Show your child how the examples in the section coincide with the questions in the practice or homework portion. For example, if your child is having difficulty with the first few questions on an assignment, those questions more than likely coincide with the first example or two in that section in the chapter. Read through the examples with your child and see if they parallel the question he is trying to solve. Use this example to help with the process of solving the problem. As always, refer to chapter 4 in this book for help with direction decoding and chapter 5 for help with the step-by-step problem-solving process and strategies.

Often, contained within each chapter, usually at the end, is a review of the chapter. This section can be particularly helpful when studying for tests and quizzes in that unit of study. Use this section as a practice test or review for your child before a test or quiz.

Another feature of many textbooks is the answers to the odd-numbered problems. These are located in the back of the book and are usually arranged by page number. If the homework is very difficult, you can check your child's progress to see if she is on the right track by checking any answers located in the back. If the homework assignment was the even-numbered problems, your child can try an odd-numbered one to see if she reaches the correct answer.

An important note is that some textbook companies also have an online copy of the textbook. This way, the book can be referred to and used for assignments at home, even if the book is at school. In addition to a copy of the textbook, these websites commonly have other resources for both parents and teachers, such as notes, online tutorials, and other reference materials. Check with your child's teacher to see if any of these support materials are available.

Note-Taking Strategies

As your child progresses through the grades in math class, at some point, the teacher will instruct the students to take notes. Counsel your child to write down everything that the teacher puts on the blackboard, overhead, computer, or visual presenter. The student should include all examples provided from the teacher as well.

These can be used as models for attempting the homework. Encourage your child to use visual pictures as much as possible. Suggest that he take all notes on the right side of the paper only; the left side can be left blank so that later at home, personal notes, comments, or questions can be added as the student reviews the notes. Then, when your child goes back to school the next day, any questions he wrote down on the left side of the paper can be posed to the teacher.

It is a good idea to use color in notes. Visually appealing notes are more enjoyable to review. If the teacher uses color in the presentation of the notes, your child can mimic this use of color. When the teacher says that some aspect of the notes is important, your child can place a highlighted box around this information, change the colors of the notes, or place stars around that segment of notes.

Homework Strategies

It may be difficult at times to assist your child with math homework. A common complaint you may hear is that you do not explain it like the teacher did or the procedure needs to look exactly as the teacher did it in class. Because you are not attending the class, have your child help you to assist her with assignments.

One way this can be done is always to have your child bring home his math book, along with any notes, papers, or guides that the teacher uses in class, as mentioned in the previous section. Have your child use a highlighter in class when the teacher is going over the assignments and have him mark questions and/or explanations in the homework that do not make sense so that you can go over them that evening at home. By working along with your child, you will have a better understanding of what is taking place in class and be more able to see the strategies and procedures as they are presented by the teacher.

Remind your child that part of math learning is making mistakes so that she does not get frustrated with a challenging assignment. The most important part of homework is the practice; if the student gets a homework problem wrong, that is okay, as long as the mistakes serve as learning aids.

When your child is solving the homework problems, this is not the time to skimp on paper. Ensure that your child is showing all steps necessary in his problem solving so that if the problem is incorrect, the child can explain what he did. Have your child leave blank space to the right of each homework problem if using loose-leaf paper. This way, when the student returns to class the next day and the teacher reads off the answers, he can ask questions about any problems that are incorrect. The space to the right of each problem is then available to write down the correct procedure as the teacher demonstrates it. Your child should not erase any work that was done incorrectly. The student can highlight the problem number and put the correct procedure to the right. It is also helpful to make written notes, such as "I added when I should have multiplied." This then becomes an instant study guide when preparing for upcoming tests and quizzes; your child can go back and review any mistakes made throughout the homework assignments and study the correct procedures.

Teach your child that working on homework is just like participating in a club or a sport. These activities have practice sessions that lead to various events, such as games, contests, or concerts. What the student does in practice helps determine how she will do in the events. How hard your child works on homework and class work will determine the progress, and eventually, the grades and success she attains in math.

Test-Taking Strategies

Many tests, especially math tests it seems, can cause quite a bit of anxiety for students. Although the concepts may have been clear the night before, your child may "blank out" on the test and literally forget that he does in fact know the concepts.

Various things can be done before, during, and after the test to help reduce or alleviate test anxiety for your child.

Before the Test

Before the test, tell your child to relax and take a deep breath. Many students get nervous when the test is at hand and need ways to relax and regain their confidence. Practice with your child taking a deep breath and counting to five to settle down and relax. Repeat this as necessary with your child until she feels comfortable doing it alone.

Often, teachers may give a review sheet or assign a textbook review assignment the night before a test. Emphasize to your child that these problems will most likely mirror questions on the test. This is the time also for proactive studying; go back through all homework assignments that are related to the test material. Key in on the highlighted homework problems—these are the problems that your child got wrong when practicing skills. By reviewing what went wrong and studying the correct procedure to the right of these problems, the student is preparing for the test.

The night before the test, and even the morning of it, give your child positive messages. Talk to the student about how proud you are of the hard work he has been doing. Send your child to school with a good attitude about the preparation and effort that's been invested. Put a note in his backpack or lunch bag as a positive reminder about the progress that's been made.

During the Test

Your child can do many things during a test to help with anxiety. One is to approach the test with an open mind. Try not to focus on what she thinks will be on the test but instead to listen or read the directions to find out what is actually on the test.

Talk with your child about blocking out other things going on. These events could be occurring in the classroom itself or could be other distractions affecting your child outside the school day. Teach your child to "put these thoughts on a shelf" to be dealt with at another time. These thoughts could be important to your child, so deeming them unimportant and just telling the student to forget them may not be the correct approach. By putting them "on a shelf," they can be put out of the way of your child's concentration and dealt with later.

Very importantly, your child should not panic. Speak with him about relaxing during the test, using positive thinking, and taking a quick mental break if needed. Just a few seconds or a minute break is sometimes what a child needs to refocus and get a fresh start on a new question.

Discuss with your child what she may face on the test. Sometimes, there is a difficult question that your child cannot solve. Encourage her to star or highlight that question and leave it alone, then move on to the next question on the test. Spending too much time on a question can lead the student to become very frustrated. This, in turn, can influence her performance on other questions. If there is time at the end of the test, the student can go back to these highlighted problems and try to solve them anew. Sometimes, another question that the student mastered on the test may give her an idea of how to conquer the difficult one.

In addition to these strategies, have your child recall the strategies and processes used when working on homework. Chapter 4: Direction Decoding covered tips for reading and interpreting the questions. It is important that these same strategies are used on test questions. For example, rephrasing a question may be what your child needs to help clarify what is being asked. In addition, Chapter 5: Little Pieces Lead to Big Problems explained in detail the strategies for tackling large problems by breaking them down into smaller, more manageable pieces. Recall these strategies with your child and encourage him to use them on the tests. These strategies will help your child develop a plan for solving the questions.

Very often, students do not see this connection between homework and assessment and sometimes need help from parents and teachers to bridge this gap. You and your child have spent a lot of time preparing for these assessments, and these skills now need to be called into action!

After the Test

After the test, help your child reward herself for a job well done. This reward may be in the form of taking time to read a favorite book or watch a favorite movie. Your child may want to call or get together with a friend, listen to music, play a game, or even take a walk.

What We Have Learned About Study Skills?

Good study skills are the primary tools necessary for learning math concepts and making them part of your child's knowledge base. Although these skills can be learned at any age, establishing good habits at an early age will give your child the best start. Set aside a time and a place to study math concepts and work on math homework each day. The amount of time spent is not as important as the routine of working on these concepts and skills.

Teach your child how to use a math textbook, both for homework assignments and as a reference. This is a skill the student will find very valuable while continuing to study math, even through the college years. Work with your child on good habits for math homework, including using notes taken in class. Remember to have your child highlight those concepts that were not clear.

Finally, have your child approach both homework and tests with a positive attitude. Teach your child the stress-busting techniques mentioned in this chapter and add some of your own! By working together through this important process, you will see your child through to a successful math experience.

Chapter 7

"When Will I Use This, Anyway?"

Part of the learning process is the awareness that what is being learned has value in real life. That is why a very natural question for your child to ask is "When will I use this, anyway?"

This chapter supplies you with some good examples of the usefulness of the math that is being taught in the classroom and the answer to the question above.

One useful practice is to take notice of all of the ways that you use math everyday and point out these experiences to your child. You, even as the parent, may be surprised at the number of times you use math, when you take the time to think about it. Some examples are when you purchase gasoline for your car, when you shop for sales, when you plan a large trip, or when you plan to buy a big-ticket item. When your child is with you at the hardware store or you are planning to paint or wallpaper a room, talk about the math needed to do these projects. Newspapers and magazines have graphs, tables, and charts that can be read and interpreted—from mortgage rates to sports statistics. Each and every day poses opportunities to be aware of the role that mathematics plays in real life.

Numbers and Numeration

Numbers and number concepts are used extensively everyday, and many times, these math concepts go unnoticed as we go about our daily routine.

Often, students have trouble finding examples of negative numbers. Savings accounts and loans are two common examples of math in real life. Money obtained from a loan, or even lunch money borrowed from a friend, are examples of negative numbers. Likewise, money in the bank or money in your pocket are examples of positive numbers. If you live in the North, temperatures are often negative in the winter months. Have your child find the difference between positive and negative temperatures to see how the temperature rises or falls.

Ratios and rates have many uses in day-to-day life. Hourly rates of pay for mowing lawns and babysitting are ratios, comparing money to time. Maybe your child wants to earn enough money to buy a special purchase; use the hourly rates to figure out how many hours he will need to work. The grocery store is a good place to practice unit rates, or unit pricing. Have your child figure out which size of peanut butter is the best buy to practice ratios. Sales at the electronic, clothing, or sporting goods store are excellent opportunities to practice percents and decimals. If you attend a sporting event, statistics, such as batting averages, are based on experimental probability, written as a decimal number to the thousandth place.

Estimation can also be explored at the grocery store. Throughout the shopping trip, have your child use rounding of prices to estimate the total bill. You could make a game out of it and see who comes closest to the correct bill amount. If you are planning a birthday party or holiday celebration, have your child be involved in planning the amount of food or prizes to purchase. You could even discuss the probability of how much of what type of soda will be chosen, based on past experiences.

Measurement

We all use measurement any time we cook or enjoy hobbies. Cooking is a great opportunity to discuss relative sizes of measuring tools, estimating volumes when refrigerating leftovers, or fractional multiplication when doubling or tripling a recipe. Sewing and quilting use a great deal of math to figure out how much material to buy and the concept of similarity with patterns that can be enlarged or shrunk to create a design.

Home repairs contribute many instances of dealing with fractional sizes of screws and drill bits, measuring lengths, and calculating areas and volumes. In planning a garden, you calculate how many plants to buy, depending on the spacing needs of the plants. Carpeting, painting, and wallpapering all involve measurement and calculation of how much material to buy.

Time management is another use for measurement in everyday life. Planning a day full of activities and using time estimates is good practice.

Geometry

Geometric shapes are everywhere we look, especially triangles and rectangles. You can use almost any situation or location to discuss geometric shapes. The triangle is the sturdiest structure; help your child to recognize the use of triangles in construction, from the bottom bracing of shelves to the many triangles found in bridge construction. Framing is an example of perimeter, and painting is an application of area. When pictures from the computer are enlarged, this is a perfect time to discuss similarity. Each of the pictures has the same images, but they are clearly different sizes; they are in proportion to each other.

Geometric shapes identify many road signs, and you can point these out to your child as you are driving. The stop sign is an octagon; a yield sign is a triangle. The double yellow lines on a road are examples of parallel lines, as are train tracks. Most often, roads intersect to form perpendicular lines.

Angles are used in sports, such as billiards and soccer. You need to understand the angles involved to drop a billiard ball into a pocket or to kick the soccer ball into the goal.

Tessellations are geometric repetitions of common shapes. There are many real-life examples of these, such as a bee's honeycomb (hexagons), the bricks on a building (rectangles), patio tiles (various shapes), soccer balls (hexagons), and quilts (squares, often). You can use these examples to point out various types of symmetry.

The game Battleship is an example of the use of coordinate geometry concepts, as are the longitude and latitude that identify locations on the globe.

Algebra

Any time you use a formula, you are using algebra; a formula is simply an equation with unknowns that have meaning in everyday life. Solving for unknown values in a formula is an example of solving an equation. For example, if you want a garden of a specific area and length, you can solve the formula *Area* = *length* × *width* to find the correct width to use. The distance formula, *Distance* = *rate* × *time*, is used when mapping out a trip in the car. You can estimate the time it will take to get to your location, if you know the distance to travel and the speed you will be driving. More advanced algebra is used in real life too, including in specific occupations such as finance, business, engineering, and health-related professions.

Statistics and Probability

Statistics is the study of numerical data. Graphs are the picture representations of data information. If you open up a newspaper or magazine on any given day, chances are very high that you will see graphs. These are good opportunities to point out the usefulness of learning how to read, interpret, and make predictions with graphs.

Your child's quarterly grade in a specific subject is an example of a mean. If there is an infant in the home, the baby growth charts from the doctor's office show a percentile ranking of height and weight.

Probability is also used in real life. Weather reports give the probability of events such as rain or snow. Lottery tickets and raffles are excellent opportunities to discuss the usefulness of learning probability. If you and your child can determine the probability of winning, you can make an informed decision as to whether you want to take a chance.

Probability and statistics are helpful in making predictions based on past events. If you are planning a big event, you can look at past experiences to see how many people attended based on the number of invitations sent out. This will enable you to predict and plan for how many people to expect. Businesses can determine how many employees to have working on given days and times, based on the probability of how busy it will be.

Smart Consumers and Informed Citizens

We live in a world that is filled with politicians and advertisers who are trying to influence our opinions. In today's culture, technology and the media play a large role in our lives. Many companies are targeting youth through television and the Internet to persuade them to try their products. A solid understanding of graphs, statistics, percents, and problem solving are crucial skills for everyone to learn to be smart consumers and informed voters in a democratic society.

When you take your child shopping and an advertisement says buy one, get one for 50 percent off, you and your child should be aware that this is not really a half-price sale. It is actually offering 25 percent off the purchase of each of two items. It would be a better deal to buy one item at 30 percent off!

Sometimes, graphs can be true but misleading. By the choice of spacing on the y-axis of a graph, a graph can be made to look very steep or very flat in an attempt to mislead. Look at the following three graphs that show the rise in number of new jobs in Adamsville.

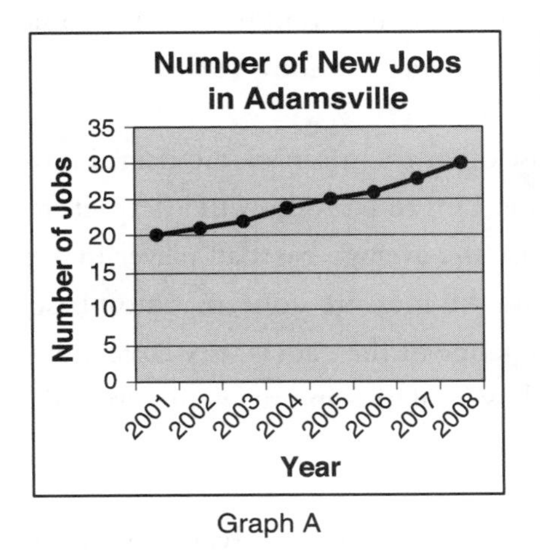

Graph A

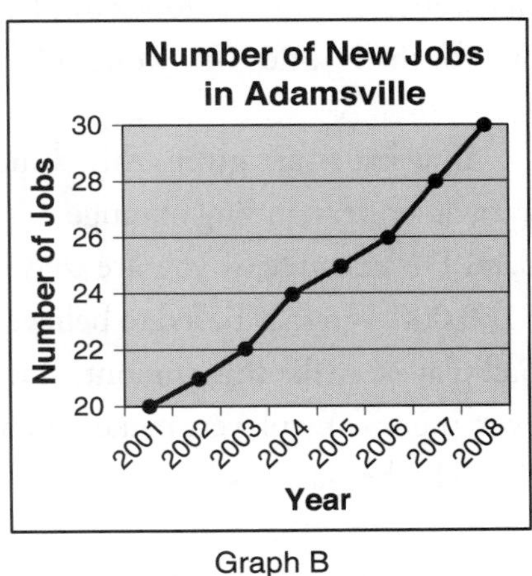

Graph B

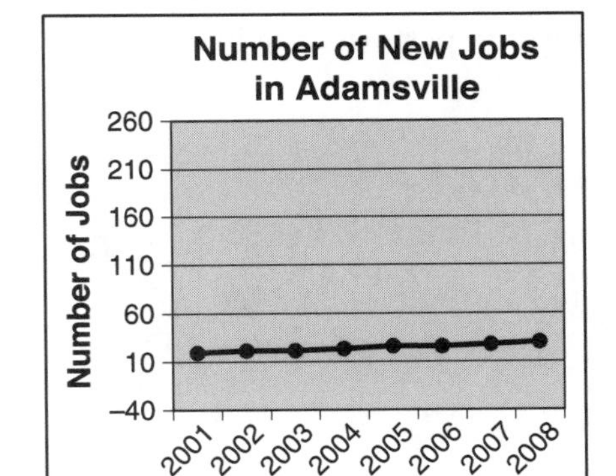

Graph C

The first graph, A, is a fairly reasonable view of the data. The second graph, B, makes it appear that there are big changes in the number of new jobs. The third graph, C, makes it appear that there is almost no change in the number of new jobs.

Often, politicians or advertisers use poll-result statistics to influence your opinion. Before you can make an opinion based on these numbers, you need to know how many people were polled and how the poll was conducted. If you wanted to know how many people would vote for a new football stadium, for example, you would not want to ask just people at a football game. This would give the poll a misleading slant.

If averages are given, this is most often the mean of the data. However, an understanding of range is important to know how to interpret the data. For example, if you are told that the average baseball player makes $160,000, you may be led to believe, on this measure alone, that most baseball players make this amount. If the range of the data is very large, however, then many players make much lower salaries and many players make much higher salaries.

As you can see in this chapter, numerous examples in real life point to the usefulness of learning mathematics. Assisting your child in recognizing this will motivate high performance in school!

Chapter 8

Parent-Teacher Communication

Parents should take an active role in the education of their children. You are, after all, your child's first teacher! Establishing and sustaining good communication with your child's classroom teacher is an important part of this process. Three words that should describe this relationship are *respect*, *patience*, and *honesty*. Many issues and concerns can be responded to in an effective manner when the parent and teacher are working together as a team.

What Every Parent Should Know and Do

A lot goes on in the classroom that parents may or may not know about. Many teachers work hard to keep parents aware of the assignments and activities taking place in the classroom, but it is also the parent's job to keep up with this information. It is common for a school, grade level, or teacher to have a "meet the teacher" night or curriculum night early in the school year. This night is a chance to get to know who the teacher is, how the class is set up, and the general expectations for the school year. It may also be an opportunity to ask questions or schedule a parent meeting with a teacher or team of teachers to discuss your child's progress.

Ask for the rules and expectations of the classroom, if they are not known. These may be posted in the room, on a handout, or even on the teacher's website. This list will be helpful in being aware of the expected behaviors and associated consequences in the class during the school year.

Pay attention to the information sent home from the teacher with your child. This communication may happen on a daily, weekly, monthly, or quarterly basis and is carefully prepared by the teacher to let you know of important events in the classroom. It may also contain family activities to do at home that make a connection with the content being presented in class. These activities can help reinforce concepts and instill confidence in your child with respect to the classroom work.

Stay informed of school activities. Read the school newsletters, be active in parent-teacher groups, and be involved in school functions in and out of the classroom.

Ask about other methods of communication that may be set up by the teacher. For example, it is common for a teacher or team of teachers to set up a homework hotline, or a phone number to call for information and/or homework. This way, the number can be used on a weekly or even daily basis to help check up on what is due in class. Another method of communication is websites that are set up with assignments, activities, and other information for students and parents. Some of these websites may also have printable worksheets posted, in case of a forgotten or lost paper. Websites often have a link to the teacher's email address. This can be a very useful way to communicate with the teacher. In addition, as mentioned in chapter 6, a copy of the textbook and related resources may be available online. Check with your child's teacher to see if this resource is available to you.

Consult your child's teacher about other resources that may be helpful. Teachers often use certain resources, whether they are books, trade magazines, websites, or so on, that are helpful. Some of these may be family-friendly and easily adapted for use at home while assisting your child. The teacher can help you select resources with activities to do with your child that coincide with what is happening in the classroom.

Most importantly, get your child involved in communicating with the teacher. The student should be the primary means of communication, and teaching your child to approach the teacher is fundamental in instilling a good rapport between them. The student should be responsible for getting notes and papers home from school. Your child should realize that she is in charge of her education and should be held accountable in ways that are age-appropriate.

When you want to communicate with the teacher about a problem, it is always best to follow a chain of command to keep a good relationship. Always approach the teacher first. If you do not get satisfaction there, then ask the teacher if talking to a counselor would help. Then approach the counselor with your concerns. The principal should be the final contact, approached only if you do not get satisfactory results from these first two resources.

What the Teacher Expects

Teachers are teachers for a reason: they enjoy teaching and working with students! Teachers want and need to be aware of things that may affect student success in the classroom. Let the teacher know if something is going on that may change a student's behavior or progress in the classroom. Sometimes letting the teacher know that there is an issue is just as important as the details. A teacher can much more effectively deal with a situation when this important information is given. If the subject is an especially sensitive one, contact a building administrator or guidance counselor to help you proceed.

Contact the teacher when there is a concern. Maybe your child had a difficult time on a particular assignment or a specific concept. Maybe your child is having a social problem in the classroom. Let the teacher know as soon as possible in the form of a note from home, an email, or phone call to school.

Remember that while the teacher tries to do his best to monitor every student every day, sometimes it is just not possible, especially with the size of today's classrooms. If the teacher does not know of any personal issues, he cannot help. Think, "How can I help the teacher help my child?"

Different teachers prefer different methods of communicating. Many like to use the telephone, and others prefer email correspondence. Some prefer to speak during the school day, and others prefer before or after school hours. By asking her preference, you are showing that you respect the teacher's time, helping to improve the relationship. Keep phone and email conversations clear, concise, and positive. Give the teacher time to respond. Listen to suggestions and keep an open mind to these strategies.

If you have a concern and your child works with a team of teachers, try to arrange a meeting with each of the teachers. In some cases, it may be possible to meet with all of the teachers at the same time. Patterns may be developing throughout the school day that all teachers are seeing and that may be important to be aware of.

At the end of a meeting with a teacher, come up with guidelines for follow-up. Who will follow up with whom, and when will this occur? Make a plan and stick to it; it may be easier for the parent to contact the teacher for an update a few days or weeks after the meeting and to check on the follow-up activities that were planned.

Teachers appreciate parents who are informed, supportive, and helpful. Emphasize your appreciation of what they do to assist your child. Try to ascertain how much help you should be providing to your child on homework and projects. Most teachers will respond positively if you solicit their opinion.

Sample Scripts

The following scenarios are samples of dialogue between parents and teachers concerning some common issues. Keep in mind some pointers when communicating with the teacher:

- Start the conversation on a positive note; give praise or appreciation to the teacher.
- Ask for the teacher's advice on how to solve the problem.
- Near the end of the conversation, restate the suggested solution to the problem or summarize the conversation.
- End the conversation on a positive note, thanking the teacher for his time.

Each of the these points will be noted in italics within the following scenarios to help model positive interaction between parent and teacher.

1. Suresh Singh is a middle school student who has had difficulty completing his math assignments recently. His mother has been trying to help him at home but is looking for assistance from the teacher. She calls the teacher, Mr. Bann, to see what help he can give.

    Start the conversation on a positive note.

    Mrs. Singh: Hello, Mr. Bann. This is Mrs. Singh, Suresh's mother. Suresh has really enjoyed your class this year, but I wanted to contact you about his difficulty in math.

    Mr. Bann: Yes, Mrs. Singh, I have recently noticed him having difficulty completing the assignments and not participating as much in class.

    Ask the teacher for suggestions.

    Mrs. Singh: Well, he seems to be having difficulty with the concepts that you are doing in class now. I have been trying to help him at home but would like some suggestions from you.

    Mr. Bann: Sure, I can help. First of all, I can meet with him during our independent work time in the morning to check over the previous night's assignment. Are you aware that the daily assignments are posted on the classroom website?

    Mrs. Singh: I was aware but hadn't used it before. I will begin using that site to check for the daily assignments and activities.

    Mr. Bann: You can also check the site for the daily notes that can be downloaded and printed out. These are the same as or similar to the ones were presented in class.

Mrs. Singh: That's great. We will do that.

Mr. Bann: I am also available after school for extra help when he needs it. Is he available to stay?

Summarize the plan.

Mrs. Singh: Yes, of course. Let's set it up for Tuesday after school, and I will pick him up. With the extra help you can give him before and after school, and my support from home checking the website and using the daily notes, there should be much improvement. Can we also follow up in a few weeks to check his progress?

Mr. Bann: That would be fine. Just contact me by email in a few weeks when you would like an update, and I would be happy to let you know how he is doing.

End the conversation on a positive note.

Mrs. Singh: Thank you so much for your time, Mr. Bann. I am sure by working together, we will see good results in the next few weeks.

2. Raquel Sanchez is a new student at Kingston Elementary School. Her family has recently relocated, and her parents are concerned with the transition. Her father contacts the teacher, Mrs. Contino, to see what the school can do to help.

   Start the conversation on a positive note.

   Mr. Sanchez: Hello, Mrs. Contino. This is Jaime Sanchez, Raquel's father. Raquel is a new student who will be starting in your class this week. She is both excited and apprehensive at starting in a new school with a new teacher. I would like to speak with you about her transition to your classroom.

Mrs. Contino: Yes, Mr. Sanchez. I have seen her name on my class list and am looking forward to meeting her.

Ask the teacher for suggestions.

Mr. Sanchez: Raquel had a lot of trouble in math at her other school, and we were wondering about the ways that you can help her.

Mrs. Contino: I would be happy to tell you. We will first set her up with a student buddy who will help her around the school and will act as a study partner. Because she has had trouble in the past with math, I will also include her in our math support class. This class is a small-group setting and will help us to determine her strengths and areas for improvement.

Mr. Sanchez: That sounds great. Do you have a list of rules and expectations we should know about?

Mrs. Contino: I do, and I will give a copy to Raquel the day she starts. It is also posted on the classroom website. In addition, at Kingston Elementary, we work in teams of teachers. Let's arrange a team meeting with the teachers who will work with Raquel, and you can come in and meet each one of us. At that time, you can get any questions answered, and we can set up a plan for Raquel.

Summarize the plan and end on a positive note.

Mr. Sanchez: It would be nice to come in and meet all of you. If you can arrange that meeting, I will speak with Raquel about her new school and what we will be doing to ease the transition. I appreciate the time you are taking to help my daughter.

Mrs. Contino: My pleasure. I will contact the guidance department and set up the meeting. We will call you with the day and time.

3. Jeremiah Chapman has been complaining at home about the amount of homework in Mrs. Wu's math class. Lately, he has not been completing the assignments, and his grades have dropped. Mrs. Chapman, Jeremiah's mother, contacts Mrs. Wu to see what can be done.

   Start on a positive note.

   Mrs. Chapman: Hello, Mrs. Wu, this is Jeremiah's mom. It was a pleasure to meet you on curriculum night, and your enthusiasm for teaching is really appreciated. I am calling with concerns about Jeremiah. He started the year doing well in math, but lately I see his grades have dropped.

   Mrs. Wu: Yes, Mrs. Chapman, I am glad that you called. Jeremiah's grades have dropped, and I think it is mainly due to a lack of homework completion.

   Ask the teacher for suggestions/keep an open mind.

   Mrs. Chapman: At home, I am hearing one side of the conversation. Jeremiah says that he does not have time in class to ask questions, so he does not understand the assignments. Therefore, he says, he can't do the homework. This is his side of the story, so I would like to hear your perspective and any suggestions you have to improve his grades.

   Ms. Wu: During class each day, there is always time for questions. Jeremiah has not been raising his hand much lately; sometimes, students are hesitant to ask questions because they think they are the only one with a difficulty. I start out the year explaining to the classes that if they have a question, chances are that five other students have the same question. By asking, they are actually helping everyone to succeed. I think I will discuss this again in class, and it would be helpful if you reinforce this idea at home.

Mrs. Chapman: That is a good point; I never thought about that. I will remind Jeremiah to ask questions in class. Thanks for the suggestion!

Ms. Wu: In addition, there is always guided practice time at the end of class where the students practice the concepts discussed that class period. This is also a time when students start their homework. I have noticed that he has not been using this time wisely. I do believe that he has the ability to do much better in the class. Do you know if he has a study hall Period 3?

Mrs. Chapman: Yes, he does.

Ms. Wu: Well, that is a time when I can answer his questions outside of class. I will give him a pass to come and see me tomorrow, and we can work together to get him on the right track. I am sure when he is feeling more confident, he will not be so hesitant to ask questions in front of the rest of the students.

Summarize the conversation/plan.

Mrs. Chapman: I agree with you. I was not aware that they have time in class to start the assignments, and will speak to him about using his time wisely in class. I will also talk with him about positive consequences for raising his grade. If you can meet with him Period 3 during school, I would really appreciate the extra attention you are giving him.

Ms. Wu: It is no problem at all. He has the knowledge; we just need to convince him of that. I will email you in two weeks with an update on his progress.

End on a positive note.

Mrs. Chapman: Thank you so much for your time and help with this. It is very important to me that he does his best.

What Have We Learned About Parent-Teacher Communication?

Parents should take an active role in the education of their child and should ask questions when they arise. Let the teacher know when there is a concern or change that could affect the student's progress in school. Use the tools that the teacher has set up to help facilitate information and communication. Keep an open mind and work together with the teacher. Sometimes you may have to take a deep breath and listen to the teacher's perspective, being open to the educator's ideas. Your child, you, and the teacher are a powerful team in the important process of educating your child!

Appendix

What Kind of Learner is Your Child?

Understanding How Your Child Processes Information Can Make or Break Academic Success

Not everyone learns in exactly the same way. Learning styles may be as unique as your child is take the following quiz with your child to discover how your child processes information; this can help you adjust studying habits for maximum effectiveness.

1. Ms. Li announces the class will be doing a new science experiment. Your child

   (a) wants to see a demonstration with pictures and visual examples of the procedure.

   (b) prefers the teacher to explain the "ins and outs" out loud as part of a class discussion.

   (c) wants to skip past the directions and learn through trial and error.

2. Mr. Rochester assigns the class a book to read. When you read to your child or your child reads independently, she

   (a) sees the details like a mini-movie or TV show in her head.

   (b) hears the narrator and characters talk aloud.

   (c) gets bored unless it's an action-thriller or mystery story.

3. It's in-school study time. To focus, your child needs to

   (a) clean his study space before beginning.

   (b) ask for quiet from his classmates.

   (c) save it for later; he works better at home, alone and outside of school.

4. During a class discussion, your child

   (a) tunes out if he has to listen for a long time.

   (b) listens to others but can't wait to respond.

   (c) listens best to classmates who move and gesture when they speak.

5. It's your child's turn to speak, and she's trying to describe something. She uses words such as

   (a) *see*, *picture*, and *imagine*.

   (b) *listen*, *hear*, and *think*.

   (c) *feel*, *touch*, and *hold*.

6. When your child gets a new toy or game that needs some assembling, he
   (a) reads all the directions and looks at the diagrams before doing anything.
   (b) asks you to read the directions aloud as he goes.
   (c) ignores the directions and just gets down to fitting pieces together.

Now, count your As, Bs, and Cs. Read below to find what your child's score says:

Your Child's Learning Style	Study Strategies
If your child chooses more As than any other letter, the learning style is *visual*. Your child understands things by picturing them.	**In Class:** Tell your child to ask the teacher to draw on the board when giving explanations. **Study time:** Keep your child's study area clean. **Terms:** Have your child close her eyes and try to see the word and visualize its meaning, particularly with math terms involving shapes or figures. **Problem solving:** Help your child imagine the setting, characters, and action as he reads. It may be helpful for him to sketch examples from word problems or to check work by graphing.
If your child chooses more Bs than any other letter, then the learning style is *auditory*. Your child absorbs new information by hearing it.	**In Class:** Tell your child to ask as many questions as needed, until she understands the lesson.

(continued on next page.)

	Study time: Find a quiet place for studying. **Terms:** Instruct your child to mentally sound out each word when learning new terms and concepts. **Problem solving:** Your child may like to read word problems aloud, giving any characters in the examples distinct voices.
If your child chooses more Cs than any other letter, then the learning style is *kinesthetic*. Your child learns by touching, feeling, and doing.	**In Class:** Make sure your child completes all in-class assignments and homework—repetition is a key learning ingredient. **Study time:** Find a private space for your child. **Terms:** Writing out the words and definitions as he uses them will help your child memorize new terms. **Problem solving:** Encourage your child to write notes, "translating" math terms into symbols or vice versa when decoding a word problem. Also, consider providing your child with study aids that can help in counting, understanding fractions, and "acting out" word problems, like little toys or snacks.

Conversion Formula Sheet

Conversions Between Standard Units

1 foot (ft) = 12 inches (in)
1 yard (yd) = 3 feet = 36 inches
1 mile (mi) = 1,760 yards = 5,280 feet

1 cup (c) = 8 fluid ounces (fl oz)
1 pint (pt) = 2 cups = 16 fluid ounces
1 quart (qt) = 2 pints = 32 fluid ounces
1 gallon (gal) = 4 quarts = 128 fluid ounces

1 pound (lb) = 16 ounces (oz)
1 ton = 2,000 pounds

Conversions Between Metric Units

1 centimeter (cm) = 10 millimeters (mm)
1 meter (m) = 100 centimeters = 1,000 millimeters
1 kilometer (km) = 1,000 meters

1 liter (l)= 1,000 milliliters(ml) = 100 centiliters (cl)
1 kiloliter (kl) = 1,000 liters = 1,000,000 milliliters

1 gram (g) = 1,000 milligrams (mg)
1 kilogram (kg) = 1,000 grams

Temperature Conversion

Fahrenheit (F) to Celsius (C):

$$C=\frac{5}{9}(F-32)$$

Celsius (C) to Fahrenheit (F):

$$F=\frac{9}{5}C+32$$

A Timeline of Mathematical Education

Guidelines for Grades 4–8 Based on the Standards Set by the National Council of Teachers of Mathematics

Note: This is *not* a curriculum, merely a summary of the path a solid mathematical foundation may be expected to take.

Over time, you can expect your child to encounter a variety of concepts under several broader categories: numbers and operations, algebra, geometry, measurement, and data analysis and probability.

Grade 4

- **Numbers and operations:** Students should be comfortable with whole numbers, decimals, fractions, and the basics of percents. They should be fluent with the mathematical operations (multiplication, division, addition, and subtraction) and be able to evaluate the "reasonableness" of their answers after choosing the appropriate solving tool for a given problem (e.g., mental math, estimation, calculator, pencil and paper, and so on).
- **Algebra:** Students should be able to solve problem situations using words, tables, and graphs. They should also be comfortable working with equations and the use of the commutative, associative, and distributive properties.
- **Geometry:** Children should be comfortable identifying the characteristics of two- and three-dimensional shapes (including location, movement, attributes, and so on) and the concepts of symmetry, congruence, similarity, sliding, flipping, and turning.

- **Measurement:** They should be familiar with length, area, perimeter, weight, volume, angle sizes, and some basic formulas to find each. Students should understand which units are appropriate for each type of measurement and know how to perform simple unit conversions.
- **Data analysis and probability:** Fourth graders should be able to make predictions before collecting data using surveys, observations, and experiments; students should learn how to represent the results in graphs or tables, analyze the results, and come to a conclusion.

Grade 5

- **Numbers and operations:** Students should be able to do problems involving fractions and decimals in real-world examples and also be comfortable translating numbers into fractions, decimals, and percents. They should be comfortable extending a number line into negative numbers and be able to describe classes of numbers according to their factors.
- **Algebra:** Children should be comfortable with geometric and numerical patterns and expressing them via words, tables, and graphs. Students should feel comfortable using operation properties (commutative, associative, and distributive), equations, and real-life models for word problems
- **Geometry:** Students should become increasingly familiar with the vocabulary of describing and classifying shapes. They should go on to learn about dividing and combining shapes and rotational symmetry. Students will also begin to work with points and lines in a coordinate graph and progress to describing, drawing, and building two- and three-dimensional representations of shapes.
- **Measurement:** A fifth-grade student should be able to estimate perimeters, areas, and volumes of irregular shapes and rectangular

solids, carry out unit conversions within a system of measurement, and understand the value of precision.

- **Data analysis and probability:** Children should be aware of the methods of data collection and representation and the effect data collection methods can have on the data itself. They should continue analyzing data sets to produce legitimate conclusions.

Grade 6

- **Numbers and operations:** Students should be proficient in comparing, simplifying, and using computations containing fractions, decimals, and percents, as well as using algorithms for greater efficiency. They should develop an understanding of ratios and proportions and feel comfortable using the properties of addition and multiplication to simplify problems and equations. The concepts of squaring and square roots will likely appear this year.
- **Algebra:** They should be comfortable building on their earlier experience with tables, graphs, and equations to begin using variables and "symbolic" equations in problem solving.
- **Geometry:** Sixth-grade students should be comfortable describing and classifying the relationships between objects using properties and precise terminology and be able to apply the principles of geometry outside the classroom in real life.
- **Measurement:** Children should be able to understand metric and customary systems of measurement and select appropriate units to measure angles, perimeter, area, surface area, and volume. They should also be able to solve problems involving rates and measurements for velocity and density.

Grade 7

- **Numbers and operations:** Seventh-grade students should be able to compare and organize fractions, decimals, and percents (including those greater than 100 and less than 1). They should be comfortable using ratios and proportions, factors, multiples, prime factorization, properties of addition and multiplication, and algorithms in problem solving.

- **Algebra:** Students should be fluent in the use of tables, graphs, words, variables, and symbolic rules and be able to identify linear and nonlinear functions in solving problems. They should be able to understand the relationships among angles, side lengths, perimeters, areas, and volumes of similar objects. Make sure your child is familiar with the transformations (e.g., flips, turns, and slides) and characteristics (e.g., size, positions, and orientations) of shapes.
- **Measurement:** Children should be able to calculate the area (or circumference) of circles, triangles, prisms, pyramids, cylinders, and other complex shapes using formulas and the appropriate units.
- **Data analysis and probability:** Students should be able to develop research and data collection strategies for analyzing populations and their characteristics, whether shared or unique.

Grade 8

- **Numbers and operations:** Students should be comfortable understanding and managing large numbers using exponential, scientific, and calculator notation. Make sure your child maintains the skills previously developed related to operations, problem-solving tools, fractions, decimals, squares and square roots, ratios, proportions, and algorithms.
- **Algebra:** Eighth graders should be still more comfortable with the properties of linear and nonlinear functions, tables, graphs, equations, variables, and symbolic algebra. New concepts will include the relationship between linear equation and their

graph, intercept and slope, and problems that require solving via graphing and analysis.

- **Geometry:** Students should be able to create proofs regarding geometric concepts, such as congruence, similarity, and the Pythagorean theorem. They should be familiar with coordinate geometry and the use of models to represent and clarify relationships between shapes.
- **Measurement:** They should be aware of what units go with what measurements and be able to apply the appropriate tools and techniques to perform calculations and find length, area, volume, angle measurements, rates, velocity, and density.
- **Data analysis and probability:** Eighth-grade students should be familiar with different types of data graphs, such as histograms, box plots, stem-and-leaf plots, and scatterplots, and measurements, such as center, speed, mean, mode, and range. They should be able to make predictions and conjectures about data results and critically evaluate the methods used in data collection.

For more information about education standards and focal points from the National Council of Teachers of Mathematics, visit their website at *www.nctm.org*.

Index

Index

Beyond Homework: Help For You and the Students in Your Life

Brushing up on the basics?

Look for Kaplan's *Outsmart* series—At home or on the go, the *Outsmart* books have the most important math, science, language arts, and history concepts students need to know. With captivating questions and amazing answers, kids age 10 & up will be ready to outsmart everyone they meet!

Outsmart History

Outsmart Language Arts

Outsmart Math

Outsmart Science

Sensing your teen is overwhelmed?

Each *SOS Guide* provides advice and encouragement for teens who are feeling the pressure from all sides. Teacher and Education.com columnist Lisa Medoff provides students with support and resources drawn from her vast experience working with teens and their families. Dr. Lisa coaches teens through difficult decisions and situations, with practical tips and suggestions for everything from communicating with teachers and parents to why it's important to get and stay motivated in school.

SOS: Stressed Out Students Guide to Dealing with Tests

SOS: Stressed Out Students Guide to Saying No to Cheating

SOS: Stressed Out Students Guide to Handling Peer Pressure

SOS: Stressed Out Students Guide to Tackling Your Homework

SOS: Stressed Out Students Guide to Getting into Clubs and Sports

SOS: Stressed Out Students Guide to Managing Your Time

Want to have fun and get a mental workout at the same time?

Try our *Brainiac* series—Each fun, portable set contains 600 flashcards loaded with interesting facts and fascinating trivia.

Challenge yourself, family, and friends to see how much you know about:

- Math, literature, science, history, and geography with the original *Brainiac Box*
- America's 50 states with *Nifty 50 States Brainiac*
- Countries and customs around the world with *Globetrotting Brainiac*
- U.S. landmarks with *History Happened Here*
- U.S. presidents and elections with *Presidential Brainiac*
- Spelling and vocabulary with *Spelling Bee Brainiac*

Does your child have the potential to compete for a top high school?

Pick up *Kaplan SSAT & ISEE*, a guide to the Secondary School Admission Test & Independent School Entrance Exam.

Kaplan SSAT & ISEE features detailed, comprehensive review for all subject areas tested, including Verbal Reasoning, Mathematics, Reading Comprehension, and Writing. The three full-length practice exams are tailored to upper, middle, and lower levels of difficulty, so your child can start preparing early and work up to the most challenging exam levels as they get older.

Is your child on the road to college?

From Here to Freshman Year

This all-in-one guide provides a roadmap to college for high schoolers and their parents, including timelines, stress reduction tips, and information about how to choose the right college. Look here for help planning courseloads, applying for scholarships, selecting summer vacation activities, managing time, and preparing for standardized exams. Then we'll help students prepare for the financial aid process, choosing when and where to study abroad, help brainstorming careers they'd like to pursue after graduation.

Worried about paying for tuition?

Kaplan Scholarships 2009 Edition features information on programs that offer significant and unrestricted scholarships combined with tips and advice on how to get them. The guide includes:

- A list of scholarships—each worth at least $1,000—that do not restrict to any one school, and do not require repayment of any kind.
- Detailed summaries on each scholarship's financial data, duration of scholarship, eligibility requirements, and application and contact information.
- Expert tips and advice on how applicants should research their options, set a timetable, apply for the best opportunities, and avoid scholarship scams.

Paying for College helps families as they begin to navigate the financial aid application process. Learn the definitions of the many terms involved in the process, research specific schools, prepare a timetable for applying, and receive the best possible aid package with our tips and resources.

Get Paid to Play features information to help high school athletes find the right level of college sports, from NCAA Division I to the NJCAA. It also offers information on scholarship opportunities as well as tips on matching students to appropriate coaches and programs. *Get Paid to Play* demystifies the process of recruiting and college applications and walks student athletes through a four-year plan that will get them where they want to be.

Look for all these titles wherever books are sold.